ON THE CUTTING EDGE

ON THE CUTTING EDGE

TALES OF A COLD WAR ENGINEER
AT THE DAWN OF THE NUCLEAR, GUIDED MISSILE, COMPUTER AND SPACE AGES

Dr. Robert F. Brodsky

Gordian Knot Books
An Imprint of Richard Altschuler & Associates, Inc.
New York

On the Cutting Edge: Tales of a Cold War Engineer at the Dawn of the Nuclear, Guided Missile, Computer and Space Ages
Copyright© 2006 by Robert F. Brodsky

For information contact the publisher,
Richard Altschuler & Associates, Inc.,
at 100 West 57th Street, New York,
NY 10019, RAltschuler@rcn.com
or (212) 397-7233.

Library of Congress Control Number:
2006927541
ISBN-10: 1-884092-62-4
ISBN-13: 978-1-884092-62-6

Gordian Knot Books is an imprint of
Richard Altschuler & Associates, Inc.

All rights reserved. No part of this publication may be reproduced, stored in a retrieval system, or transmitted, in any form or by any means, electronic, mechanical, photocopying, recording, or otherwise, without the prior written permission of Richard Altschuler & Associates, Inc.

Cover and book design: Bette Brodsky

Printed in the United States of America

Distributed by University of Nebraska Press

TABLE of CONTENTS

ACKNOWLEDGMENTS ➤ ix
PREFACE ➤ 1
INTRODUCTION ➤ 4

CHAPTER 1 GRAD STUDENT DAYS: 1947-50 ➤ 6

    The Making of a Rocket Scientist ➤ 8
    WW2 interrupts college, discharge leads to career options

    Great Court Cases, #1 ➤ 10
    In which we meet Orville Wright and testify testily

    Dragon Slaying 690 ➤ 15
    ENIAC, Smeniac—a first start in computational fluid dynamics

CHAPTER 2 THE ATOMIC YEARS: 1950-56 ➤ 20

    What! No Flying Saucers?! ➤ 21
    A new start in the Wild West!

    Fun in the Pacific ➤ 23
    A-Bomb testing and viewing at Eniwetok

    Straighten Up and Fly Right ➤ 28
    Making unruly bombs hit the side of a barn door

    Mexico Under Siege ➤ 34
    In which an unarmed TX-5 bomb is dropped on our good neighbor

    Angle Iron to the Rescue ➤ 43
    Fixing the first externally carried A-Bomb, the Mark 7

    Mick Takes a Dare ➤ 50
    An insouciant Navy Commander breaks the sonic barrier

The A-Bomb Enables the Space Age ➢ 56
The space age began in New Mexico – the Bomb got us there

A "17" for Uncle Joe ➢ 60
The H-Bomb was our answer to Stalin's A-Bomb arsenal

CHAPTER 3   GUIDED MISSILE DAYS: 1956-58 ➢ 68

California, Here I Come! ➢ 69
Getting into missilery on the ground floor

A Wonderful Early Simulation ➢ 70
The analog computer, in its day, performed miracles

The Glitch at Mach 2 ➢ 74
A lesson in hardware inspection is learned

CHAPTER 4   THE FABULOUS YEARS AT AEROJET: 1958-71 ➢ 80

The Roaring '60s ➢ 81
The Space Age enters like a lion

Flatass Jack of Jackass Flats ➢ 82
Moving the huge Saturn 2nd stage from Downey to the sea

Who Knew? ➢ 88
Getting into the spacecraft business from scratch

"FIRST" in Space ➢ 92
30 years too soon on a rescue lifeboat for the Space Station

The Walking Wheelchair ➢ 100
From Moon walker to kid's stair walker—a real conversion

The Sea Bee Saga ➢ 105
Sea launch the easy way

Dirty French Postcards ➢ 110
Peddling an automated syphilis tester in Western Europe

The Algerian Galitzianer ➢ 114
Shady business dealings with a wily, mean CEO

The Road to Maroc ➤ 118
Building a Comsat ground station in the middle of nowhere

Caveat Emptor ➤ 122
A young engineer saves the nation's most critical space program

CHAPTER 5   FREEZING IN IOWA: 1971-80 ➤ 128

Down on the Farm ➤ 129
Living there, after they've seen Paree

The Iceberg Cometh ➤ 130
An international iceberg conference in landlocked Ames, Iowa?!

Pie in the Sky ➤ 135
The butt of David Brinkley's evening news and SPAM's chagrin

Great Court Cases, #2 ➤ 140
Or, Annals of Goofball Engineering: A Twin Beech crashes

In 1984, I "Invented" Astronautics ➤ 144
How astronautics became an accredited curriculum

Ode to Teaching ➤ 151
A paean to the second of the world's oldest professions

CHAPTER 6   BACK IN OUR OWN BACKYARD – TRW: 1980-88 ➤ 156

Return to Industry ➤ 157
Back to California; to Industry and Academia

The Shuttle Bus Caper ➤ 159
A failure in judgment by a slow learner

Beating the Fog Factor ➤ 164
Winning the Mars Observer study and Gene Spangler's heart

Good Cop, Bad Cop ➤ 168
A slapstick team runs Independent Research & Development

Old Cyberspace U. ➤ 172
Fight on USC—The modern way to bring education to the masses

CHAPTER 7   THE TWILIGHT ZONE: 1988-2004 ➤ 176

    No Way Retired! ➤ 177
    A retirement of a sort—with windmill tilts

    NIH (Not Invented Here)—The Crew Return Vehicle Metamorphoses ➤ 178
    40 years later—the renewed fight for a space escape lifeboat

    The Sweet Surveillant Science ➤ 182
    The amazing capabilities of our "eyes in the sky"

    The Hydrogen Economy ➤ 186
    The frustrating fight to establish a new world order

    A Man Without a Company ➤ 191
    After 60 years, an alumnus of no one!

CHAPTER 8   SUMMING UP: 2004-2005 ➤ 194

    Letting Go—Gradually ➤ 195
    Musing on a career. Should I have been a real doctor?

    The Glory of Engineering! ➤ 196
    Even if it's done badly, it's a hell of a ride!

    The Last Hurrah ➤ 201
    Old sailors and engineers never die—they travel!

    The Great Events of the Last Century ➤ 206
    A review of major aerospace achievements after the Wright brothers

INDEX ➤ 214

ACKNOWLEDGMENTS

In my youth, I was encouraged to pursue aeronautical engineering as a career by my parents, Samuel and Sylvia Brodsky, and my dear nurse, Rose Steck, who lived with us until I went away to college. "Smilin' Jack" and "Buck Rogers" in the Sunday funnies fanned my interest in aviation and outer space. Family and friends regaled me with aviation type books, and I still retain some wonderful ones from the early 1930s, such as Grover Loening's *Our Wings Grow Faster* and Le Corbusier's *Board the Airliner*. I made gas model airplanes, which my father and I chased in his car all over Montgomery County in the northeast Philadelphia area.

Sylvia and Sam Brodsky at the Stork Club

And most of all, seeing that I was not interested in the family business, my parents encouraged me to pursue aeronautical engineering as a career.

My first attempts at non-technical writing took place at Iowa State University in the '70s, where the germ of the idea for this book was formed. Since my retirement from teaching at USC in 1996, I have been writing fulltime, in addition to sailing twice a week and attending the South Bay Writer's Workshop once a week.

And I've been helped: I thank all my collaborators for their insights and the joys of recalling the old times. A noted author, BH Friedman, a Cornell classmate, has continually given me guidance and always reminded me to "keep your day job." One new friend and former crewman, Mike Kerrigan, himself a writer, helped me both passively and actively. Another crew member and former boss at TRW, Bob Walquist, gave great aid and comfort in punctuation and the like. And, of course, Patti, my dear wife of almost 50 years, has read and commented on every page. She has been especially helpful in telling me when I get too technical or when I relate more detail than she wants to know. Finally, my Writer's Workshop group has been extremely helpful in making suggestions that turned dull stories into brighter ones, and always toned down my pomposity.

But the people I truly owe the biggest debt to are the many great men and women who were my comrades and coworkers in my many ventures. We fought, we played, we fretted together—and generally came out feeling good about what we achieved. Living and dead, they know who they are.

The book is dedicated to my wife, Pat; our children (especially Bette, who has always been a great fan and who designed this book), and grandchildren; my extended Brodsky family, of which I am presently the patriarch; and all our friends who have made life so interesting—and such a ball! And, finally, I want to thank my editor/publisher, Richard Altschuler, for guiding me in its production.

PREFACE

When I was young—very young—I fell under the spell of aviation. I dreamt of being a pilot—dog fighting as in World War 1—with a white scarf around my neck. I dreamed dreams of flying and going to the stars. I never expected that most of my dreams would be realized.

My 60-year journey through the high-tech world of aerospace has been a wild ride. The chronologically ordered 8 chapters of this book include stories that illustrate the achievements as well as the fun and foibles of this journey. It begins in graduate school, following my discharge from the Navy at the end of World War 2, and then moves into a strange world governed by the AEC, the U.S. Atomic Energy Commission, which was engaged in a crucial hurry-up effort to establish an arsenal of atomic weapons as our main chip in the Cold War. The action here took place mostly in Albuquerque, New Mexico at Sandia Corporation, and at the nearby Los Alamos Scientific Laboratory. The progress in nuclear weaponry made during my 6-plus years, from approximately 1950-1956, at Sandia is no less than remarkable, has not been reported elsewhere, and has clearly changed the world in which we now live.

Hearing the clarion call of the upcoming age of guided missilery, I next moved to Southern California ("LaLa Land") to learn a new trade. I became Chief of Aerodynamics at the Convair/Pomona plant, where I went to help develop the Navy fleet's Defensive Weapons array. These were the ship-board fired guided missiles: Advanced Terrier and Tartar, and then the shoulder fired antiaircraft missile called "Red Eye." But then, suddenly and miraculously, the Space Age cometh and calleth—my dreams were coming true. It heralded the start of the golden age of the '60s, where wondrous projects and deeds occurred, culminating with the Apollo program.

Thus began a pioneering 13 years of life drama and workplace shifts—from California to Paris, France, under the aegis of the Aerojet and Space-General

Corporations of Azusa and El Monte, California. This stint encompassed the "borning" of our space effort, as I became involved with launch vehicles, experimental rockets, moon walkers, biological warfare, and early satellites and spy spacecraft —always struggling with trying things that had never been tried before. And I got to run Aerojet's office in Paris. Me, an ex-Seaman First Class gets to advise and pow-wow with NATO Generals and high order government muckety-mucks!

When the space industry almost shut down, at the end of the Apollo era, I felt that most of the mountains had been climbed, and sought to fulfill an early desire: to become a teacher. So, by an unusual set of coincidences, I found myself in Ames, Iowa, as a Professor of Aerospace Engineering and head of the Department. Being in the world of aerospace academia and away from the industry for the first time proved to be a major challenge; I soon found out that great teachers are born, not made. Nevertheless, I managed to be the first to develop new courses, for both the classroom and laboratory, in space technology, which eventually led to the fairly recent establishment of "Astronautics" as a legitimate and degreed academic pursuit. The 9 years in the Midwest went by quickly. A late '70s sabbatical leave from Iowa State, solving problems at Hughes Aircraft close by the LA Airport, proved to be exhilarating, and a portent of things to come.

I found that I missed the excitement and competition of industry. My family and I also realized how much we missed Southern California and year-round sailing, and how nice the South Bay area was. Returning to Hughes would be a conflict of interest on their part, since they had now created an annual "Visiting Professor Program" —so pleased were they by their experience in my just-completed year. Hiring me away from academia would make the newly founded program look like a recruitment ploy. I scratched around and found a good "Senior Systems Engineer" job at their arch rivals, the equally noted and close-by TRW. I gave up tenure at Iowa State and started working there in August 1980.

At TRW, I mostly worked on proposals for new hi-tech projects, and directed the Space and Technology Group's research efforts, retiring in 1988 at age 63. I also had the opportunity to resume teaching my space courses at night at the University of Southern California, from 1982 until 1996, when I retired from USC. I continued active consulting in the business, and started writing—which I do to this day in 2005, the time of this writing.

The aim of this book is to give the reader some insight into this high-tech, cutting edge world, as seen first hand by the author. It provides vignettes of progress made by my fellow workers, and gives insight into technology so new and different that the "blind" were literally leading the "blind." The stories I tell herein are of a semi-technical nature, but are easily understandable for technical and non-technical persons alike. Read on and see what it was like to be "On the Cutting Edge."

Sketch of an imaginary rocket ship by my son, Robert Derby Brodsky.

INTRODUCTION

The great majority of the GIs who had returned to college after World War 2 had one sole objective: Get it out of the way, and get on with lives that had been interrupted by the war. So, when I returned to Cornell to complete my senior year, my attitude was no longer that of just trying to get by—as my life had been pre-Navy. I was a bear for knowledge and got straight A's, in spite of running down to Greenwich Village on weekends to hear or play the hot jazz music that I loved.

After graduation from Cornell in May, 1947, I moved back to a rent-controlled Greenwich Village studio apartment, at 108 W. 12th Street, at $8 a week, and played regularly in a band on Sullivan Street. As the fall semester approached, I had to decide whether to go to graduate school or really try to make it as a jazz musician. On the one hand, New York University had a degree program in Aeronautical Engineering, a field that had always attracted me. I still had G.I. Bill money available, and I could make a deal for additional support working for a professor. I played most of the summer, and continued this 9 p.m.—4 a.m. routine weekdays, after school started in October. In the first year of grad school, I was a fulltime student, taking courses from about 4 in the afternoon until 7 or 8 at night, generally at the downtown Washington Square campus near both my apartment and the band venue. During the early morning, I slept for a while, (after our gigs, we generally tapered off by going to my room and talking, playing my fabulous 78 rpm records, drinking beer, and smoking until around 6 a.m.) then did homework—which continued on the IRT to the uptown engineering campus in the Bronx. There I proofread the encyclopedia that my mentor was writing, for a handsome 10 cents per page fee, and also attended an early class.

In my second and third years, major life style changes took place. I began teaching 2 classes, one a lecture, the other a laboratory class. I played only on rare weekday occasions, when the kid who had ousted me couldn't make it. The teaching was a blast! The students were all so avid and so questioning that this must surely have been the golden age of teaching. During this period, I found time to get married and have a baby daughter. As a result, I gave up flying, which I had continued to do even after I busted out of the Navy V-5 flight training program, because I simply could not consistently land my advanced trainer safely. We moved uptown to more sedate surroundings. I was settling down to a life in engineering and academia, though left with the question I still have: Could I have made it in the world of the old time, funky jazz I continue to love so much?

The 3 years at NYU went by quickly, including the adventures covered in Chapter 1. As completion of my doctoral thesis neared, I started looking for gainful employment. As you will see, fate—in the form of flying saucer visions—landed me in the "Wild West": Albuquerque, New Mexico, no less!

Sandia Corp. parking lot outside of the first engineering building, circa 1949-50. Manzano/Sandia Mountains in the background. (Albuquerque, New Mexico)

Chapter 1
GRAD STUDENT DAYS
1947-50

When I started my graduate studies at New York University, the noted Dr. Alexander Klemin had already retired as chairman of the Aeronautical Engineering Department, but still maintained an office on the campus. Among his many activities, he began compiling an enclyclopedia of aeronautics. In 1947, he hired me as his chief proof reader.

Klemin (1888-1950) was born in London, May 15, and immigrated to the United States in 1914. He became a U.S. citizen while head of the Aeronautics Department at the Massachusetts Institute of Technology. During World War I, he was the officer-in-charge of the research department, Army Air Service, at McCook Field, Dayton, Ohio and then was head of the Guggenheim School of Aeronautics at the NYU College of Engineering from 1925 until he stepped down in 1945. He was the author of several text books, including the *Encyclopedia Of Aeronautical Engineering*, *If You Want To Fly*, *Simplified Aerodynamics*, and *Airplane Stress Analysis*. For a while, he was my hero!

THE MAKING OF A ROCKET SCIENTIST

In the beginning, space engineers were made, not born. You had to bring some fundamental knowledge to the show, but early space work was largely empirical—you felt your way. The stories to follow will illustrate the arduous path to space, and the some of the quirks of fate that got us there. Let's start in late 1941:

I was reared in Philadelphia, the only son of well-to-do Jewish parents. I went to an historic all-boys high school, Central High. My Father and I were listening to the radio on Sunday morning, December 7, when FDR broke in. I was 3 weeks from high school graduation, and had been accepted at Cornell and MIT. I chose the Mechanical Engineering curriculum at Cornell. Having "skipped" 3 half-years of elementary school, I started college in February, at age 16. Because Cornell was now on a war-time basis of 3 semesters per year, I had almost completed my junior year when the Selective Service Board sent for me. They said they doubted that my student deferment from military service would be continued after the completion of the next semester. If you did nothing, you would probably end up in the infantry, I did something! I joined the Navy in the V-5 Flight Program in Ithaca in early 1944, having just soloed in a Piper Cub as a civilian.

I was mustered out over 2 years later in May 1946, no worse for wear and tear. I had busted out of advanced flight school and ended up out-of-college and into boot camp. Although I was generally a screw-up in the service, I did nothing to harm our war effort. I literally grew up during my time in the Navy, from a wise-guy kid to a wise-guy adult, with a small touch of maturity. I got an "early" discharge for the ostensible reason of resuming my college education in June, which I actually did resume in the fall. My discharge was expedited by my taking a wild chance and telling a yeoman who ran discharges that I was, indeed, a cousin of his good friend, Joe Brodsky of Newark, who was to be married 4 days hence. Of course, I had never heard of Joe, and knew of no family branch in New Jersey. Thinking quickly, I told his friend, the yeoman, that I really would like to go to Cousin Joe's wedding. "You got it, Sailor!" I was out in two days! The post-war adventure into the world of high-tech begins.

At boot camp in 1944; then as a grizzled veteran 6 months later.

GREAT COURT CASES, #1

As soon as the cross-examination began, I sensed the drops of cold sweat from under my arms running down the sides of my chest. Unstudent-like, I had dressed for the occasion with white shirt and tie, and a blue blazer sports jacket covering what must have been an ever enlarging wet spot in the shirt underarm area. The cross-examiner was vicious. His job was to discredit me in the eyes of the jury for my part in the ongoing suit action. It was like child's play for him to demonstrate to the jury that I was far from an expert witness. After he established my status as a callow youth, his first zinger came, "And how many times in your 22 years have you been involved in an airplane crash investigation?" Then, "And I understand that it was you who took the photographs at the accident scene?" Next, two zingers, "And do you consider yourself an expert photographer? What are your qualifications?" Finally, a long discussion keyed on his question, "And how did you know where to take the pictures, since you have told us that you are neither an expert photographer nor a crash expert?" Mercifully, after I fearfully mumbled the deprecatory answers the wicked prosecutor was seeking, the attack was over. The friendly lawyer's redirect quietly but forcefully brought out that I was merely acting under the directions of his principal true expert witness, the re-nowned Dr. Alexander Klemin, my employer and ranking doyen of the burgeoning helicopter field.

I first met Professor Klemin in late summer of 1947, while I was being edged out of my job as cornetist, by a superior musician, in a jazz band playing in a bistro on Sullivan Street, off of Washington Square, in the Village. I then began to think in terms of going to graduate school in the Aeronautical Engineering Department of New York University. Dr. Klemin had been its Head for many years prior to his recently reaching the mandatory step-down age for department heads. I still had considerable eligibility left from the G.I. Bill which would not only pay my tuition but also around $100/month maintenance. This was precisely the same amount I was earning as a musician, except for the free food and drinks. He was impressed by the fact that I had been a reporter, sports editor, and, in my just completed senior year, associate editor of the Cornell Daily Sun. He said that if I would enroll in grad school, he would hire me, at 10 cents per page, to proofread his ongoing Encyclopedia of Aeronautics. Moreover, after the first year, if I was doing well, he would get me a job as a lecturer in the Mechanical Engineering Department, since my undergrad degree was in Mechanical Engineering.

Dr. Klemin was a brusque, crotchety guy, with great swings of comradery and aloofness. He reminded me of Sheridan Whiteside, the "Man who Came to Dinner." At this time in life, Klemin had serious trouble with his legs, a condition I now think was due to spinal stenosis. He needed at least one wrist-gripping crutch to move around, and sometimes used two. He was a bear for work and loved the helicopter business. He rejoiced in the success of his prize student, Frank Piasecki, who now had his own business, based on his unique design concepts. When Frank came to visit the Heights campus, he was a walking role model, and these visits were celebrated by everything short of a parade! Although "Doc" Klemin could drive, he didn't like to, and depended more and more on me to chauffeur him around, as well as do other 'gofer' tasks. To his everlasting pinchpenny credit, though, he afforded me the opportunity to meet the great men of early aviation.

One day in late 1947 he said to me, "Mr. Brodsky, tonight I'm going to the Wings Club and would like you to take me there, and be my guest for dinner." Wow! I had heard of the club, of course, but never in my wildest dreams did I ever expect to go there. It had been established in the 20s, when the New York City area was still a hotbed of aviation. I surmised that it was like a man's drinking, eating and discussing sanctuary for the early and earliest of birdmen. On that first visit there, unbelievably, I briefly shook hands with Orville Wright (who died a few months after), Glenn Martin, and Grover Loening! Even though these encounters were perfunctory, the elation of even being in the same room with them sticks to me forever. At a later evening, I met and talked at some length with Leroy Grumman, whom I knew to be a fellow Cornellian. He told me to look him up after I got my graduate degree (at that time I only had my sights on a Master's degree in Aero), and he would see about getting me a job. To this day, I can't believe how lucky I was to have these tremendous brushes with legends. I think Klemin did this for me out of shame, for the pitiful wage he was paying me —but what a pay-off!

There I was – shaking their hands!! Orville Wright, Grover Loening, Leroy Grumman — the early birdmen. A TWA DC3 is in the back ground.

My wretched experience on the witness stand came about as result of a fatal crash of a Transcontinental & Western Airline DC-3, which was blamed on an oddball freak of weather. On a winter day, it had ploughed into dormant farmland at full bore, some 3 miles short of its intended LaGuardia Field runway, killing all of the 15 passengers and crew. The ordinary course of events following such a tragedy was that, by an earlier general agreement between the airlines and the government, the insurance company would pay the heirs of the deceased a flat $5,000 per fatality. No successful suit against such a settlement protocol had been made to date. In this case, however, some prominent power structure passengers were among the deceased, and their families pressed what would now be called a 'class action' suit to try to obtain a more realistic settlement. Believing that big money might be an ultimate outcome, a first line NYC litigating law firm agreed to argue the case.

The hiring of Dr. Klemin as their chief expert technology witness was a foregone conclusion. He was at once a true expert, a noted and revered figure, and readily accessible—being the Head of the only Aeronautical Engineering Department in the vicinity. The requisite legal proof necessary to have any chance of beating the airline's defenders, Lloyds of London, as I recall, was to prove negligence on TWA's part. Nothing else, such as the even-then ubiquitous "pilot error" dodge could win the day. Dr. Klemin was unable to visit the crash site because of his relative immobility. However, on examination of crash site pictures in the newspapers, his intuition led him to the probable cause of the crash. He thought he detected a difference between the ground scars made by the left and right propellers, and suspected that when the plane hit the ground, only one engine was operating.

If he could prove this, our side could rightfully claim negligence and righteously sue for a bundle. He said to me, "Mr. Brodsky, I want you to take my car to the crash site and take a lot of pictures of the furrows made by the propellers after the very first ground contact." He supplied me with a very good camera and ran me through some camera practice sessions at the nearby playing fields. He also provided me with the proper credentials to gain access to the crash site.

I brought back a bunch of winners that made him break out in a broad smile, as he carefully reviewed each picture. He saw exactly what he expected! For the first time, I won a commendation from him. He informed the law firm of his "gotcha," anxiously awaited his day in court, and asked me to drive and accompany him to the downtown courtroom. When the time came, he laboriously (using both crutches) worked his way into the witness box and, using blow-ups of the most revealing pictures, brilliantly made a devastating case for airline negligence. The minute he finished his testimony, I felt it was all over.

Now approached the wicked cross-examiner! "Dr. Klemin," he began, "I notice that you have some difficulty in getting around. Were you really able to take those pictures, which are at the foundation of the case, by yourself? The crash site appears to be somewhat inaccessible and far from a road." Klemin replied testily that "no, he hadn't been to the site," but that he had instructed his trusted graduate assistant to do the job. "Oh," the lawyer queried, desperately taking a stab at a long shot, "And is that person in the court today?" On getting an affirmative answer, he immediately excused Klemin, probably feeling he could only do his case a disservice by further probing, and waited for the first legal opportunity to get me sworn in and up on the stand. But, despite his utter destruction of me as a credible witness, the facts of the case could not be denied.

For the first time in history, a very large airplane crash judgment was levied, and I had simultaneously aged 5 years in the longest fifteen minutes I ever remember!

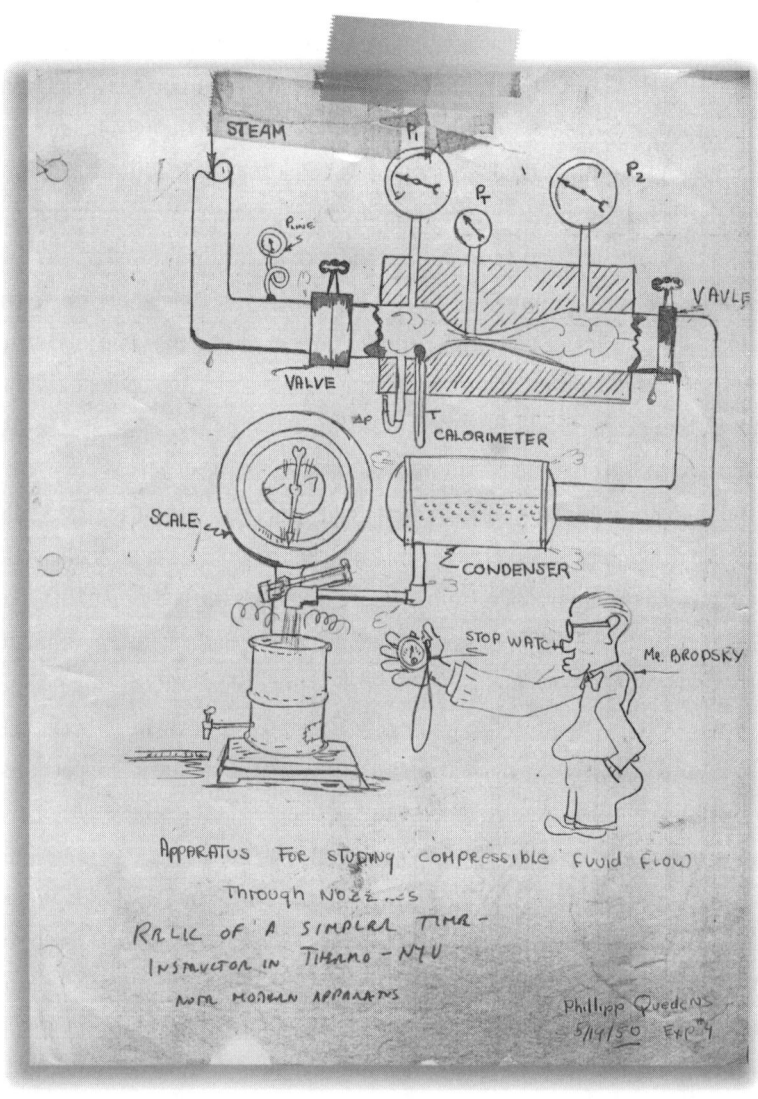

My job as an instructor in the mechanical engineering department consisted of teaching courses in thermodynamics and conducting laboratory courses.

DRAGON SLAYING 690

Since my major course at Cornell was in Mechanical Engineering, I had to take some basic undergraduate courses in Aeronautics in order to get up to speed. I carried 3-4 courses per semester; not a hard load. I earned my master's degree in one calendar year, and was invited to continue on. There being no work readily available in industry, it was my best option.

My doctoral advisor, Dr. C.T. Wang, had received his degree from Brown University and was a structural dynamics expert. He believed that some of the analytical methods used by structures people could be applied to fluid flow problems. In doing so, he was encroaching on the academic territory of his department's fluid mechanics expert, Dr. J.F. Ludloff, which led to their emnity.

For this experiment, Dr. Wang guided me and a fellow graduate student, Socrates de los Santos, whose famous naval father was on a Phillipine postage stamp. In later life, Dr. "Soc" ran the Naval Ordnance Lab's large supersonic wind tunnel in White Oak, Maryland. Unfortunately, Dr. Wang became a victim of cancer and died young, after publishing two excellent texts.

The dreaded calamity, a computational error, which all doctoral candidates fear, was discovered about two months before I planned to have my thesis completed, and be ready for the final oral defense of my work. I had a wife and a new baby to feed. My wife had quit her job to raise our daughter, leaving us a one-salary family. I dearly needed to have a real job, after many "poor" months on the G.I. Bill dole, NYU Instructor's salary, and parents' help.

I knew that the computational error I had just stumbled upon would have minimal effect on my final calculations and conclusions. But I faced a moral dilemma: Should I tell my advisor, Dr. C.T. Wang, about it, or forever keep it to myself? I had been working on the calculation for over 6 months, painstakingly punching a Marchant mechanical calculator, which was the fastest, most accurate finger-actuated calculator in 1949. I was trying to depict mathematically the high speed air flow around a 2-dimensional lifting body, a task heretofore possible only by using data derived from wind tunnel testing. I had a humongous set of interrelated equations to solve, hence the need for so much calculation time. I was finally beginning to see the end. One day, to my utter horror, I found that I had early-on made a mistake. I knew that the error would continue to propagate in all later computations. Should I junk 6 months of effort? "No, I won't," I said to myself, "The error will be of minor importance and no one but me will ever know about it."

In making this decision, I thought about the fate of a fellow candidate under Dr. Wang's aegis, one Socrates De Los Santos. A couple of months ago, Soc had carped angrily to our mutual advisor, "What do you think I am, a coolie?" This after the good doctor (and he was a good and brilliant gentleman, and a great teacher) asked him to do more work, after he was sure he had done enough to finish his thesis. This gaffe, coming at a time when we were essentially on the same graduation schedule, cost Soc exactly one year in his quest for the doctorate. I could not afford to have a similar fate befall me.

But the secret error preyed on my mind and gave me a lot of angst. I decided it would be dishonest to ignore, and might come back to haunt me the rest of my life. Finally, I damned the torpedoes and told Dr. Wang of my troubles, fully expecting to be reamed out for wasting his money and ordering me to start over. His actual reaction was diametrically opposite from what I expected! He smiled and said, "Ah, Mr. Brodsky, what we have here is an opportunity I've been looking for. The University of Pennsylvania has just sent me an announcement that they are seeking scientists to use their new 'high speed' digital computer, the ENIAC."

He explained that the ENIAC was reputed to be the world's first practical electronic computer, and he naturally wanted to know more about its operation, since his research always involved a lot of computations. Because my parents lived in Philadelphia, he knew that sending me there would not be a financial hardship. He said he would arrange for my getting both instructions on programming as well as machine time. In exchange, he required a full detailed account of the ENIAC operation and its working environment upon my return. So my honesty, albeit delayed, gave me the opportunity of a lifetime.

By telephone, he arranged for me to access the ENIAC over a month's time span, mostly—being low man on the totem pole—in the dead of the night. The machine was a horror of early electronic design. Racks and racks of vacuum tube-filled equipment took up two large primitively cooled rooms. Knobs, pushbuttons, and dials abounded. Wire cabling was strewn all over the floor. Technicians, ready to chase troubles and replace tubes, stood by around the clock. It truly was the modern equivalent of a sweat shop. When I first saw it, it scared the hell out of me. Could this bucket of bolts really save my life?

I bunked and ate at my parents' in Germantown, and bussed, trained, trolleyed, and sometimes—on call, in the middle of the night—taxied to the campus. I told my mother how apprehensive I was about ever being able to conquer this beast. In typical fashion, she said, "ENIAC, smeniac, *abi gezunt*!" The latter was a Yiddish phrase literally meaning "As long as you're healthy," but usually applied in such famous sophisms as "Cancer, smantzer, *abi gezunt*! With this wisdom to back me, I bravely carried on. I was trained on the fly during the day, and soon learned how to prepare inputs to the machine. In less than two weeks, I had learned how to program and could now prepare my inputs.

Miraculously, in 3 non-consecutive nights of operating, I completed the work that had taken over 6 months on the Marchant! Not only that, but the results looked reasonable. I was then able to finish my thesis, was back on my hoped-for schedule, and ready to accept a job I had interviewed for. As I write this, I realize that what I, and the other pioneer ENIAC toilers at that time, had accomplished marked the very beginning of what is now called "Computational Fluid Dynamics" (CFD), wherein the flow around very complicated bodies, such as the Space Shuttle, can be calculated in velocity regimes and under atmospheric conditions that can not be obtained in a wind tunnel. Of course, while calculators with the power of the ENIAC now fit in the palm of your hand, the CFD supercomputers revert back towards its larger size.

Returning to New York with my calculations completed, I easily and successfully defended my thesis work. I was aided greatly by a vicious drawn-out verbal battle between my advisor and his great Germanic rival on the faculty, Dr. Ludloff, an aerodynamicist of considerable skills, who hated Dr. Wang for muscling in on his territory. Their heated repartee began immediately after I tried to field the third question that Dr. "L," hoping to discredit his enemy's student, threw at me. After the battle subsided, they were both so winded and apoplectic as to give short shrift to further questions. They soon announced that I had passed the orals, having shown remarkable acumen.

Now you know why they call the PhD, "Piled Higher and Deeper." And, perhaps, there is a moral here. By slaying the dragon of academic dishonesty, I reaped a just reward.

The ENIAC —the world's first practical computer: Housed at the University of Pennsylvania.

The ENIAC, left, was the first electronic computer. Now, engineers are trying to squeeze spacecraft systems onto a tiny chip, right.

California Classroom

A Learning Link to the Space Place

Imagine you were alive in July 1969 when the Apollo 11 spacecraft landed on the moon. To learn more about the moon or space exploration, you would have gone to the local library to check out books to read. You could not point and click to something on the World Wide Web because it wasn't there yet. In fact, nobody had computers in their homes. Computers were usually found in government offices. They were so big that one computer filled an entire room! Compare that with today's laptop computer. It has more memory and is more powerful than all the computers that commanded Apollo 11 to the moon.

Imagine you wanted to call your friends to tell them about the moon landing. You would have picked up a telephone with an old-fashioned dial instead of today's push buttons. Touch-tones were not in wide use yet. In fact, there were no other kinds of telephones: neither cordless nor cellular. If your friend wasn't at home, you would have to call back later because there were no answering machines to leave messages on.

Most of the extras we use today, like ATMs, VCRs, cable television, CDs, hand-held computer games and beepers, didn't exist three decades ago. This is because the computer chips needed to make all these machines run hadn't been invented yet. But since 1969, computer engineers have come up with ways of making computers much better, more powerful and a lot smaller.

They have created methods of squeezing more and more information and memory into smaller and smaller computer chips.

Despite the primitive computer technology in 1969, we were able to land a man on the moon.

Today, 31 years later, we send spacecraft to planets billions of miles away. Imagine what we will be able to do in another 30 years!

The Space Place is a Web site for children. For fun facts and activities about computers and other space missions, visit http://spaceplace.jpl.nasa.gov/x2000do2.htm. This article was provided by NASA's Jet Propulsion Laboratory, managed by Caltech in Pasadena.

A year 2000 celebration of the ENIAC that appeared in the Los Angeles Times.

Chapter 2
THE ATOMIC YEARS
1950-55

"WHAT! NO FLYING SAUCERS?!"

Almost at the outset of the interview meeting, Pete Petersen asked, "Do you have any problems about working on a highly secret government project? Sorry, I can't tell you any more." My ears perked up, my spine tingled, I inwardly giggled; I kept a poker face. This was it! I knew it in my bones! I had hit the mother lode! These suckers were making flying saucers and I was getting in on the ground floor!

By late '48, it appeared probable I would finish my doctoral degree work within a year. I found a blind ad for "advanced government work in Albuquerque" in the New York Times. The company involved was Sandia Corporation. Nobody, faculty and friends alike, had ever heard of them. But I had a prescient hunch. I applied and was asked to come for an interview. For almost a year, reports of almost daily sightings of UFOs in the Southwest had been flooding the newspapers. In fact, because of their frequency, sighting reports were already being relegated to other than the front pages. "Could this be the Sandia action?" I mused.

The interview with Dr. Petersen and other obviously high supervisors only served to strengthen my feeling that some mighty mysterious things were going on here in this desert outpost of about 36,000 souls. And, unlike my native New York born-and-bred schoolmates, who did not recognize that there was any real life west of the Hudson River, I could seriously consider their verbal job offer and not be judged insane.

The formal offer that followed—a miraculous $450 per month with a one-month vacation per year—seemed too good to be true. With family assent, I agreed to become about the 700th employee. When I left in 1956, 6 plus years later, Sandia had well over 5,000 employees, and Albuquerque's population was over 200,000 and still growing. Yes, I admit to being disappointed when I found that they were making atomic bombs instead of UFOs, but it soon became very exciting, and being surrounded by PhD-laden physicists provided me with an outstanding engineering and management growth environment. I took advantage.

Sandia recruitment—circa 1952. Boomtown, USA: two years earlier, when I arrived on the scene, Sandia had less than 1,000 employees. There seemed to be one bulldozer for every 10 people.

FUN IN THE PACIFIC

In Spanish, Sandia means watermelon, and that was the normal color of the Sandia mountains guarding the eastern edge of Albuquerque at sunset. Sandia base was adjacent to the next southernmost range, the Manzano mountains.

The mission of Sandia Corporation was then to design, build and test atomic bombs. The lab was first operated by the Western Electric Company, with leadership later transferred, in the early '50s, to their research arm, the Bell Telephone Laboratories.

The atomic warheads were designed and built by LASL—the Los Alamos Scientific Laboratory. Drop testing of inert test bombs took place at the AEC's Salton Sea base in southern California, as well as at Lake Bemidji in Minnesota, where extreme cold weather bomb drop testing took place.

Live-bomb testing took place in the Pacific, and later, for smaller yield bombs, in Nevada. The testing of dummy bombs for flying characteristics, trajectory determination, and fusing and firing ability was the responsibility of Sandia Corporation. It was supported by the 4925th Strategic Bomb Squadron stationed at abutting Kirtland Air Force Base.

In 1949-50, I had just arrived on the scene.

Dressed in traditional garb—Hawaiian shirt, short pants and sun-shielding straw hat—I was on board a Navy landing ship, standing next to Mel Merritt, about 65 miles from the expected "ground zero" drop zone. The tension was mounting, as the multitudes of task force personnel on our and other ships awaited the detonation of the first test A-Bomb to be delivered by airdrop. The viewers consisted of scientists, engineers, military personnel and civilian contractors. Our particular work had been accomplished in the past two-plus weeks. We hoped that the hundreds of pressure-measuring instruments we had placed on our test structures would all read out correctly. We missed our compatriot, Jim, who was not allowed to come, lest his newly operated-on ears be blown out by noise or shock wave passage.

When I first arrived on the job, I was given a non-secret, on-loan assignment as a stopgap, until my super-secret "Q" security clearance came though. I sat in an office with two physicists, Drs. Mel Merritt and Jim Shreve, who worked in the Weapons Effects group. Mel was a very intense, serious person, who did not have even the slightest sense of humor that I could detect. But he was an excellent teacher and experimentalist, and obviously a big brain. I immediately named him (behind his back) "Young Dr. Malone," after the star of a steamy soap opera that was then popular. Jim, on the other hand, was very outgoing socially and lots of fun, until he turned off his hearing aid, and settled down for his day's analytical work. A little later, when the now-famous Lovelace Clinic, then across the street from our office, perfected its ear rebuilding operation, Jim was persuaded to give it a try, to correct his congenital, mechanical, inner ear problem. The operation was a complete success, but it nearly ruined his life. He simply could not concentrate at work, because of the background noise. After a while, he got earplugs and reverted back to his normal mode of silent, happy operation, removing the plugs only at lunch to socialize.

Mel and Jim were getting ready for an A-Bomb test in the Pacific. Their assignment was to design tests to determine the destructive radius of an atomic blast against targets such as houses, buildings and factories. Their approach was to measure pressure distributions on the surfaces of such targets, and then to have structural engineers find out the probable degree of destruction, by either analysis or loading tests. In a relatively small corner of Sandia's huge "back lot," hard by the Manzano's, they had established a high-explosive TNT firing range, to test instrumented, scaled models of typical structures. In getting ready for the forthcoming Pacific tests, they were simultaneously preparing full-scale models of the same targets, for use in the Pacific. Mel was "instrumenting" both scale and full-sized models with fast response pressure pick-ups, which would record and transmit before they were destroyed by the shock wave. The full-scale building models either were to be placed on atolls or, for the most part, on barges, anchored in place at different positions relative to ground zero.

Jim was trying to calculate analytically the interaction of a high pressure shock wave—such as the bomb created—on both odd-size individual structures and rows of structures, like the models we were having built. We did not yet have direct access to the few new-fangled computers based on the ENIAC design. One, in Washington, was Sandia's property, but it had been commandeered by the Army, and was strictly devoted to making bomb trajectory tables for the first generation "production" atomic bombs, soon to come on line. Los Alamos had the third extant large computer, but it was used full time for warhead design. Without a machine to carry out the higher order calculations governing the shock wave flow, Jim's task was a very difficult and slow one. He had to get by on a hand-punched Marchant mechanical computer for the time being, until Sandia's onsite machine arrived, so his progress was marginal.

My assignment turned out to be more amenable to fairly simple slide rule calculations. I was assigned to determine pressure distributions around structures far from ground zero, where the shock waves were very weak (unlike Jim's very strong shocks), and consequently where structure destruction might or might not take place. This work was "right down my alley," and I loved the mathematics it involved. My doctor's thesis work had depended on my finding analytical solutions to mathematical formulations of higher order—and I just "ate up" that kind of challenge. I found that a simple wave theory, formulated by the British physicist Sutherland, could be applied and excellent solutions found quickly, without the need for a high-powered machine. I was in my element—happy—and soon willing to forget that I was hired on "false pretenses." All the more so, since my paper solutions of pressure distribution predictions were all closely borne out by Mel's TNT, instrumented, range test results.

Because of my continued analytical pressure distribution success with scale model buildings, of ever more complexity and reality, I became the local "fair-haired boy." The payoff came almost simultaneously with the award of my special "Q" clearance. Before I started my pre-ordained assignment, which was to aerodynamically design the bomb shapes and tail fins, I was invited to help Mel set up and witness the test off of Eniwetok atoll. Jim was not allowed to go because of his recent ear operation. The live-bomb test was to be an airdrop, and a whole fleet of ships, target barges and ancient disposable ships, aircraft, and instrumented buoys were involved. It was a massive under-taking. Mel and I were responsible for setting up and checking out the pressure measuring instrumentation on the barge-carried and atoll located model city segments we had ordered up.

By now, I had my uniform-of-the-day Hawaiian shirt and was ready to go. We got to the Marshall Islands via Military Air Transport, whose undying tradition was to "break down" in Honolulu for a couple of days coming and going. This was my first South Pacific adventure. Waikiki Beach was everything people had said it would be, but the anticipation of going further westward towards Eniwetok detracted from my enjoyment. Once at our final destination, we set up our office in military barracks. From among the veritable army of scientists, engineers, Sea Bees, ships of the fleet, security forces and workers that swarmed over the island, a team was assigned to assist us. We had about two weeks to accomplish a small miracle, and by "working our asses off" we made it on schedule, with only about 15 of our hundreds of pressure transducers inoperative.

We observed the bomb drop from the deck of an LST, about 65 miles from the projected ground zero. The drop aircraft was a B-29 from the 4925th Squadron. We were all equipped with a thick piece of opaque material, to shield our eyes from the blast, and were warned that we would be told when we could remove it, to see the anticipated mushroom cloud. I saw the light of the blast through the thick screen, and felt the heat simultaneously; the former was muted by the eye shields, but was still spectacular, and the latter was much hotter than I would have expected at that great distance. When we were given the OK to remove the eye shields, the sky scene was unbelievable. The angry black smog, boiling as if from a witch's cauldron, spread rapidly upwards. The shock wave thunderclaps first arrived more than 5 minutes after the heat pulse, almost as an afterthought. The whole show was awesome. I felt that the demon's fury had been unleashed. It also made me understand why a common Sandia Base occurrence was the arrival once or twice a month of a padded wagon, whose attendants would sadly lead away a distraught veteran engineer or scientist.

What remains burned in my memory was the sheer magnificence of the post-ignition vista. The hideously angry sky contrasted with the calm, almost slow motion, scene on the ocean's surface. Here, one could see the shock wave moving over the surface, easily toppling the boats and ships as they were engulfed. I had visions of a mighty Thor wielding his awesome hammer. While watching, I had a ridiculous flashback: Did I really hear right when some-one told me that the granting of my "Q" clearance had been delayed two months while the FBI thoroughly investigated a youthful peccadillo—when my errant baseball broke a cranky neighbors window? How did that relate to this?

Later, when my two older children asked me, "Daddy, how could you have done such work?", I gave the standard answer of all people who, like myself, could live comfortably in the A-Bomb business: "Somebody had to do it, and it sure was interesting work." I held the thought, and still do, that if we hadn't dropped the bomb, my glorious naval career would have been greatly extended. It would even have been possible for me to have been sent—following the trail blazed earlier by my dear Link Trainer instructor school friend, Mary Margaret Maureen McMichael—to the Pacific theater, thereby diluting the glory of my coveted and hard won "Zone of the Interior" service medallion.

View from our LST. It was the first detonation from an air-dropped test bomb. Previous tests in the Pacific were conducted from towers on land. Among the ships that you see in the picture were our barges with our instrumented simulated buildings on them. The LST (landing ship tank) carried the sightseers. Courtesy of Sandia National Laboratories

STRAIGHTEN UP AND FLY RIGHT

After the "Fat Man" fell on Nagasaki, the government decided that we needed an arsenal of high energy, or "high yield," atomic bombs. It formed the Atomic Energy Commission (AEC) to oversee and run the show; formally established the Los Alamos Scientific Laboratory, under the aegis of the University of California/Berkeley, to continue research and development on warheads; and asked the telephone equipment manufacturer, Western Electric, whose Bell Laboratories subsidiary was newly famous for developing the Nike ground-to-air defense missile, to found and operate the new Sandia Corporation in Albuquerque, whose job it would be to engineer, produce and test the bombs.

The first bomb exploded in anger over Hiroshima, named "Little Boy," was a gun-type weapon. It assembled a critical mass of nuclear material, by actually shooting a half load of the critical material down a massive gun barrel, in order to mate with the other half, and thus achieve a critical mass of nuclear material. Such bomb types are low-tech, and consequently almost certain to work. But they are inherently limited to a low yield, since they explode before the two pieces of nuclear material can be sufficiently squeezed together to realize their full potential. However, because they were so ruggedly built, they later found use in bombs designed to penetrate either the earth or dams, and explode only after they had achieved such penetration. The first of such production A-Bombs, the Mark 8, was developed by the Navy as a collaborative effort between the Naval Ordnance Test Station/Inyokern (California) and Sandia and Los Alamos.

The desire for higher yield bombs—hundreds of kilotons of equivalent TNT destructive power instead of tens of kilotons—led to the simultaneous development of the more risky implosion type bomb, as exemplified by the "Fat Man."

The "Trinity" test near Alamagordo, New Mexico, had earlier proven the principle, and Nagasaki proved out the technology. In an implosion bomb, a spherical non-critical mass of material is squeezed down until it achieves the density necessary to sustain a nuclear chain reaction. This is done by surrounding the nuclear core by a sphere composed of many individual, snuggly-fitting segments of high explosives, which are designed to focus the ingoing blast wave front, so as to uniformly squeeze the nuclear core until criticality is achieved. The TNT explosions must be initiated simultaneously in the many outermost surface segments around the periphery to obtain the desired high yields. This had to be done with great skill to assure simultaneity of blast wave initiation all around the sphere. The resulting inward traveling process is called "implosion," as opposed to an outgoing "explosion." The larger the diameter of the high explosive segment layers, the more the TNT, and the higher the yield due to more "squeezing" of the core material.

The Mark 3 was a pre-production, one-of-a-kind prototype bomb. It assisted in later version internal and structural design, but was abandoned because of production-related considerations. Just before I arrived on the scene, the Mark 4, the first bomb designated for the nuclear arsenal, had been designed by my predecessors, all physicists, in committee-like fashion. As might be expected, it turned out to be a "camel." There was trouble—serious trouble! It couldn't fly the same flight path, even if dropped under identical conditions. It was not possible to make bombing tables used by bombardiers, to plan and execute bomb release, that were worth the paper they were printed on! How did this happen? And how did my group, now composed chiefly of aerodynamicists, save the day?

The rules of the game in the aerodynamic design of the original family of A-Bombs were simple: They must fit in the existing rectangular bomb bays of both the B-50 bomber and its soon-to-be-available replacements, the B-36 (which had two identical, tandem bomb bays) and the B-52; the maximum diameter of the bomb must be as large as possible to get maximum yield; and the bomb could have no moving parts, for reliability reasons. The latter restriction was a "killer," because it meant that the length of the tail fins needed to stabilize the bomb could not extend beyond the rectangular bomb bay envelope. The 4 short stubby fins had to be aligned in a catty-corner fashion at the end of the bomb bay.

With such restrictions, a blind man or a neophyte just out of college, like me, could have designed the bomb's shape without too much thought: put the warhead

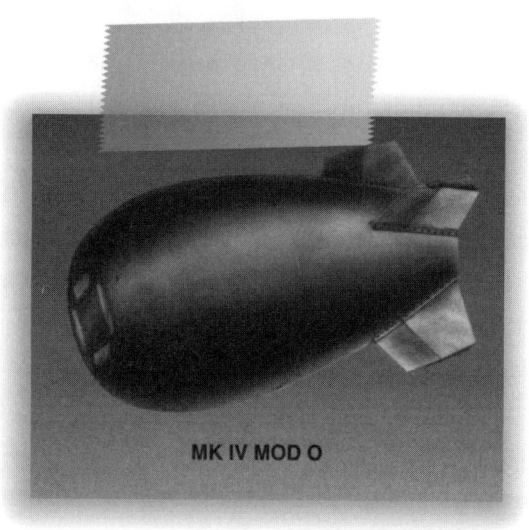

The general shape of the early production bombs

sphere as far forward as possible, so that the center of gravity would be forward, thus easing the stability problem, since the tail fins were going to be ridiculously small in span. This dictated a flat nose, which, for streamlining purposes, could be almost the size of the warhead radius. Aft of the maximum diameter section, the bomb skin should taper back with a conical after-body to a flat base, whose diameter could be smaller than the warhead diameter. That would allow the 4 stabilizing tail fins, like the tail feathers of a dart, mounted at the end of the after-body, to be as far aft of the bomb's center of gravity as the envelope would allow. What you get looks like the body shape of the MK 4 MOD 0 shown in the picture.

The only truly open major design option was the shape and size of the tail fins themselves. Some of the best aeronautical minds in the country were pressed into service to help answer the question of finding the optimum airfoil shape of the pitifully small tail fins to achieve the maximum stability. Sandia, having no resident aeronautical engineers on board at that time—after all, I was the first to be hired with such training—but with the clout of a giant, had commandeered an aerodynamics panel to give advice on demand. This panel consisted of the top aerodynamicists and aircraft designers in the country. On their first meeting, they suggested that Sandia, overly populated with physicists, hire some aerodynamicists. Apparently I was the proverbial "Johnny-on-the-Spot."

On the panel were Jack Northrop, founder of Northrop Aircraft; George Schairer, Chief of Aerodynamics at Boeing; Ira Abbott, a legendary engineer, who was Director of Research at NACA, the predecessor of NASA; Ed Heinemann, Chief Engineer of Douglas Aircraft/El Segundo, Paul Hill, chief of NACA's Pilotless Aircraft Research Division; and Dr. Alex Charters, who ran a very long, partially evacuated, gun range at the Aberdeen Proving Grounds in Maryland, where up to 1-inch diameter models could attain transonic speeds—at a time before the advent of transonic wind tunnels. Other great men were later added to the panel as the need for their specialized advice became apparent.

The panel had met in Albuquerque, before my appearance on the scene, and had discussed, but did not sketch, proposed plan forms for the tail fins. The panel had suggested using a "double wedge" fin cross section. By this, they meant a diamond-shaped cross section, with both sharp leading and trailing edges, such as were being used in the designs of the just developing X-series of transonic aircraft. Alas, the non-aero Sandians interpreted "double wedge" to mean a cross section that began as a diamond shape, but starting at the middle continued to extend rearward at an even greater angle, leading to a broad-based trailing edge, instead of a sharply pointed one. It turned out that this misinterpretation was a very fortuitous one, since the high drag caused by the blunt based fins helped a great deal in the stabilization process. With all that advice and snafus, the Mark 4 shape was born and ready for wind tunnel tests—where the trouble first revealed itself.

The blunt tail end of the bomb design allowed a very novel method of testing bomb stability. We placed a thin rod, called a "sting," into the open aft end of the 6-inch diameter models; and this rod attached to a ball bearing located at the center of gravity of the bomb. With this type of mounting, the Mk 4-shaped bombs could oscillate about 8 degrees off the centerline before the rod hit the back end of the model opening. With proper instrumentation, we could measure the change of bomb angle relative to the sting, as we increased the speed of the air in the tunnel up to the limit of over nine-tenths of the speed of sound. The full-scale Mark 4, when dropped from high altitude, would actually reach speeds only slightly higher than this speed, Mach 0.90, so the simulation was a good one.

We did our wind tunnel testing in one of 3 large, high-speed tunnels. The choice depended on availability and cost at the time. My favorite was the fabled, but now defunct, 12-foot Southern California Cooperative Wind Tunnel, located in Pasadena off of South Raymond Street, near the still-present power plant. Not so fond are my memories of the other two tunnels we used: the one in cold, freezing Buffalo, operated by the Cornell Aero Lab and directed by the improbably named King Bird; and the 10-foot tunnel at Wright Field, in equally drab Dayton, Ohio, run by—need I say it—Zip Zonars. But all tunnels showed the same thing: when the Mach number approached the speed of sound, the bomb oscillated like crazy, sometimes hitting the stops. The problem was that the full-scale bombs did the same thing during drop tests, and the oscillations were serious enough to change the trajectory. The worse problem was that we did not know how to predict whether a bomb would oscillate or not, since we believed an unpredictable wind shear encountered on the way down was needed to set off the wobbling motion.

WIND TUNNEL SET UP FOR CWT, CAL, & WRIGHT FIELD
THE 6 INCH DIAMETER MODEL COULD SWIVEL IN PITCH ~ PLUS OR MINUS 8 DEGREES.
NOTE: REFLECTED SHOCK WAVES NEVER HIT REAR END OF MODEL

- ROTARY BEARING
- MODEL STING (HOLDER)
- MACH NO. < ~.92
- STRONG TRANSONIC SHOCK WAVE
- REFLECTED SHOCK WAVE
- 10 OR 12 '
- STING SUPPORT
- NO SHOCK WAVES BELOW ~ M = .87 OR .88

So the Mark 4 literally couldn't hit the "side of a barn door," but, nevertheless, it went into the arsenal—warts and all—waiting until my group figured out some tricks to reduce the size of the oscillations as a retrofit. As warhead technology and our "fixes" improved, the approximately 20-kiloton-yield Mark 4, essentially retaining its basic exterior shape and tail fins, evolved into the vastly more powerful Mark 6 and Mark 13 versions. The latter had yields well over 100 kilotons, and was the precursor to the coming H-Bomb.

The fixes we tried were all empirical, and some worked better than others. All tried to prevent or restrict the movement of the transonic bow and body shock waves, which we believed exacerbated the oscillations. Even today, over 50 years later, I hesitate to go into more detail for fear of breaching security. The fun and excitement in the testing was in seeing how the different "fixes" performed. We were able to reduce the full-scale oscillation from about 10 to 13 degrees to 3 to 4 degrees. This was small enough to have a negligible effect on the trajectory and allowed our computer operators to confidently make good bombing tables.

Later in life, when I was seeking employment, my reputation as "fixer" of dynamic stability problems won me a Chief of Aerodynamics position at a missile company whose main product exhibited control problems at a crucial point in its flight. Little did they know of the "voodoo nature" of our solutions at Sandia!

MEXICO UNDER SIEGE

When I returned from the atom bomb test in the Pacific, I resumed my regular job—the one I had been hired for—as the first trained aerodynamicist at Sandia. My new group had on hand a polyglot mix of civilians and military personnel. Ken Erickson wanted to reorganize this melange by technical discipline, thereby hopefully alleviating the chaos. After a while, he reassigned our boss, George Hansche, since he was not trained in aerodynamics, and asked me to take over the group. Emboldened by this promotion, which included a salary increase, we decided we were in New Mexico to stay. We gave up our rental at 405 N. Adams and, with my father's down payment help, and bought a bigger house across the street at 410 N. We soon made friends with our new next door neighbors, Colonel Osmund J. and Jean Ritland. He would play an important part in my career, both at Sandia and later in California when, as a Lt. General, he commanded the ballistic missile program from the Los Angeles Air Force Base in El Segundo, down the street from my job at TRW.

My new group's first assignment was to do the aerodynamic design of the TX-5, which, after development and testing, was destined to become the second line of A-Bombs to go into production, and into our newly growing atomic arsenal, as the MK 5.

Once started, the countdown—ten!, nine!, eight!... —proceeded normally to "release!" But then, nothing happened! Through the high-powered tracking telescopes we watched the A-Bomb, obviously loose from its shackles, rattle around in the bomb bay. The plane's crew must have known there was a problem, for we saw the contrails jiggling side to side, as if the aircraft were emulating a dog trying to shake off fleas. It seemed like an eternity before the pilot started a left turn to avoid flying over Mexican air space. Only then did the bomb reluctantly begin its earthward free fall. I heard Colonel Os shout, "Ohmigod, it's going to hit Mexico!" The enormity of his yell hit me like a ton of bricks. We were atomic bombing a friendly neighbor!

The pre-drop flight path of the silvery gleaming Boeing B-47 was southward down the middle of Salton Sea, heading a bit west of Mexicali. It was a gorgeous fall morning, with clean, crisp air and totally clear skies. Inside the bomb bay was the TX 5, the first A-Bomb whose aerodynamic design and flying characteristics my neophyte group was completely responsible for. A main purpose of the drop test of this inert prototype bomb was to see if we had, indeed, tamed the seemingly random pitching oscillations that plagued the first production bomb, the MK 4. We used the same airflow spoilers that we had found worked in our wind tunnel tests of both the Mark 5 and the similar Mark 4/6 small-scale models. These "fixes" reduced the random oscillations to a very tractable 2-3 degrees in the tunnel—a performance that had not yet been proven in full scale. I personally had a lot riding on its success—it was my new group's "baptism under fire." Sandia and the AEC had recently established the highly sophisticated instrumented bombing range at the southeast end of the Sea. It had the ability to accurately track the drop trajectory and receive radioed information of bomb firing functions and oscillations. The AEC had, for security reasons, decreed that all pieces of test bombs must be painstakingly collected after land impact, to preclude giving up any design secrets. Since the bed of the Salton Sea was a gooey morass, with apparently no known bottom, bombs dropped in it did not need to be recovered. The same was true of the less sophisticated Lake Bemidji bombing range in Minnesota, used for extremely cold weather A-Bomb drop testing.

The MK 5 bomb, whose still experimental TX version was being drop-tested that day for the first time, was designed specifically for carriage in the Air Force's newest tactical bomber, the 6-engined Boeing B-47. The first one of these sleek, powerful, graceful beauties accepted by the USAF was assigned to Colonel Ritland's 4925th Strategic Bomb Squadron. The Colonel, who had joined me and several others to witness the test, manned an observer's tracking telescope next to mine. These so-called "phototheodolites" had very large optics, making it possible to see small objects at huge distances, so long as we kept the incoming aircraft inside the viewing screen, by correctly cranking the horizontal and vertical telescope pointing actuators. The loudspeaker informed us of both the probable arrival time of the aircraft and coordinates to use, to aim our telescopes so as to locate the plane as it came into sight at around 35,000 feet.

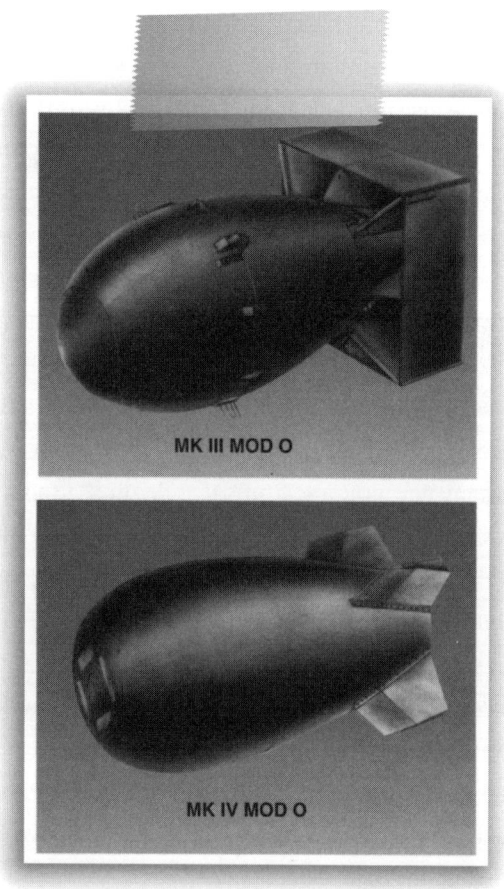

Early strategic A-Bombs. The MK 3, top, was a later version of the "Fat Man" —the Nagasaki bomb. The MK 4 (mod 1) was the first "production" bomb in our ever-increasing arsenal. The MK 4, MK5, MK6, and MK 13 all had similar shapes.

Almost simultaneous with the failure of the bomb to emerge from the bomb bay after the count-down reached "zero," the base emergency klaxons deafened the air, and an armada of Army, Navy, and Air Force vehicles, their sirens screeching, stormed off towards the Mexican border, in a confused sequence that seemed right out of the Keystone Kops. Trucks and crews of all varieties streamed toward the now forewarned border, with the mission to pick up and crate every piece of splattered junk they could find. In one fell swoop, we had caused a serious security breech and created a major diplomatic incident, attacking—fortunately—a barren piece of Mexican desert land that was located some 5 miles south of the border!

Although the bomb was unarmed, we needed to convince the Mexican government that we were just fooling around, and really meant them no harm, coupled with the solemn promise never to do it again. Unbeknownst then to me, the telephone lines between Sandia, Washington, and Mexico City were ringing off the hook! When I returned home, I soon found that the Air Force continued to be in a mild state of panic. The reason was obvious: they either had or soon would have a larger number of the brand new in-air, refueled B-47's than B-52's in their inventory; and this force combination—B-47 with MK 5—represented our major strategic deterrence at this pre-ballistic missile point in our history.

They hurriedly convened what I soon found out was a new Air Force "invention"—a so-called "Tiger Team"—whose purpose was to find a quick solution to a big problem. Each of the principal organizations were asked to name two delegates to the team. On the Sandia team were the MK 5 Program Manager, Eaton Draper, and myself; on the Air Force team were Col. Os and the pilot who had flown the ill-fated test mission. Boeing supplied Thornton A. ("T") Wilson, their B-47 Program Manager, and Boeing's Chief Engineer, George Schairer. He also was a member of our "Aerodynamics Panel"—an expert advisory group that the AEC had assembled to help us neophytes. As fate would have it, "T"—an Iowa State University Aerospace Engineering Department alumni—in his later position in the '70s, as CEO of Boeing, would be my benefactor, while I was head of his former department there.

Before the first Tiger Team deliberation, I assembled some of my group, including Alan Pope, a key idea man, Paul Rowe, and others to kick around ideas about the phenomenon and how to fix it. We quickly almost unanimously arrived at what was the most probable reason for the bomb's reluctance to leave the bomb bay upon release. We opined that some vagary of air circulation in the open bay provided enough lift to cause the bomb to "float" in the confined area into which it was crammed. Enough lift must been produced to counteract gravity, thus preventing the bomb from dropping.

Thornton A. "T" Wilson, Jr. T became the department's most distinguished alumnus by rising through the ranks to become the Chairman and Chief Executive Officer of the Boeing Company. He was honored by Iowa State with a rare honorary doctorate in 1993, and in 1998 he and his wife endowed a chair in Engineering at ISU with a $1 million endowment. T died in 1999.

LOS ANGELES TIMES

Thornton Wilson; Boeing Chief Cut Costs, Created Economy Jets

Below: The sleek Boeing B-47 at first roll-out. It was developed specifically to carry the MK5 bomb, which was smaller than the earliest bombs that required larger air- craft, like the B-50 and B-36 to carry them. Courtesy of the Boeing Company

By ELAINE WOO
TIMES STAFF WRITER

Thornton A. Wilson, the former chairman and CEO of Boeing whose radical cost-cutting kept the aircraft company afloat amid the gas shortage and deregulation of the 1970s, died in his sleep at his Palm Springs home and was found by family members Saturday morning.

He was 78 and was credited with developing the fuel-efficient Boeing 757 and 767, the aircraft that helped the company dominate the commercial jetliner market.

Wilson took over Boeing as president in 1968 when it was near collapse and administered bitter medicine to keep it alive. He slashed 95,000 jobs, reducing the work force by almost two-thirds, and kept the payroll lean until profits began to climb.

"I was a demanding and tough manager," he said years later. "I was no joy to behold."

His strategy for Boeing also included diversifying, so the company began to study alternative energy and transportation, producing light-rail systems for Bos-

the B-47 swept-wing bomber, then became overall project engineer of the B-52 program during the later stages of its development. He also led the proposal team that won Boeing the contract to produce the Minuteman intercontinental ballistic missile.

In the 1960s, Wilson began a swift rise through management ranks, becoming vice president in 1963 and executive vice president in 1966. Soon after he became president two years later, Boeing was reeling from an almost disastrous overestimation of the initial market for the 747 jumbo jets. The company had no commercial orders for the craft for three years. Wilson's cost-cutting caused much bitterness in Seattle, where Boeing was the largest employer.

Although the company's finances were still precarious, in the late 1970s Wilson gambled $3 billion on developing a new line of fuel-saving jets. Boeing had to risk churning out new models, he told Fortune magazine, just "to stay in this business."

The results of the gamble were the 767, rolled out in 1981, and the 757, in 1982.

executive officer in board in 1972. He when he retired. Collier Trophy, the honor, for pushing jetliners. He was Aviation Hall of National Business

his wife, Grace Sarah Parkinson III and Daniel dren.

To prove this theory, Alan suggested a novel wind tunnel test approach, which was made even more palatable by the fact that the wind tunnel in Pasadena we used could reasonably simulate both the speed and drop altitude conditions of the full scale test at Salton Sea. He proposed that we mount a model of the B-47 underbelly, with bomb bay and open door geometry duplicated in scale, on the top ceiling of the test area section of the tunnel. We would measure the vertical forces on a model bomb in the simulated bomb bay by placing a wind tunnel balance inside of it, supported by a rod coming out of the aft end of the bomb. If the vertical forces recorded by the balance were near zero, then our point would be proven! We estimated that we could be ready to do this in a week, since we already had a model and balance system on hand; and Ken Crowder, our model designer/builder, said he could build a partial B-47 model post haste, if Boeing supplied the needed drawings. We further presumed that the Air Force could readily muster the priority to bump any present wind tunnel occupant, since A-Bomb work was alone at the top of anybody's priority list.

I brought these ideas to the initial Tiger Team meeting a couple of days later, and our theory was "bought" on the spot! The team then speculated about quick fixes. The obvious one—to somehow positively push the bomb out of the bay by a hydraulic or explosively operated ram—was deemed premature. Ossie Ritland said that such a displacement bomb rack was under development by Ed Heinemann's Douglas/El Segundo Aero Group, but that it was at least 6 months away from testing. It was meant to assure the safe release of bombs and other ordnance carried under the wing or fuselage of attack aircraft. These armaments sometimes had the nasty habit of flying up and bumping into the carrier aircraft after they were released. Colonel Os speculated that what might work for these so-called "external store" armaments should also help in our case.

Because of the urgency of the situation, we tried to concoct a faster "fix." We soon arrived at the idea of breaking up the circulatory airflow in the open bomb bay by effectively dividing it into two bays. This could be done by placing an inverted U-shape, cutout baffle around the bomb at its midway point. The baffle would be attached to the bay exterior structure and stay with the plane after bomb release. This might be an easy and inexpensive fix until the new positive displacement bomb rack came along. We also urged the Air Force to place higher priority on the Douglas program.

At this juncture, Colonel Os jumped up and said, "We can build a baffle tomorrow, and we'll run a Salton Sea drop test the next day, if a test bomb is available! We'll make the damn baffle from some of the 3/4-inch plywood I've got in my garage." It turned out that his unbridled enthusiasm was thwarted by the necessities of having Boeing design, build, and install the baffle while simultaneously having the Air Force B-47 Project Office at Wright Field in Dayton, Ohio, responsible for "blessing" all aircraft modifications, sprinkle its brand of holy water on the scheme.

But, my neighbor's impulsive action made an everlasting positive impression on me. This was my first view of a real "can do" manager with the guts to make things happen, and I admired him forever for his bravado. As life went on, I found that such people were few and far between, but that those who took the chance and were successful inevitably ended up as "President."

Because of the red tape delay, we now not only had time to conduct the static wind tunnel test program that we had formulated prior to the first Tiger Team meeting, but also to plan a second wind tunnel test, which the ever-ingenious Alan Pope had proposed. I don't know whether his approach had ever been done before —but it certainly was news to me and the rest of our ad hoc bunch. Alan suggested releasing model bombs from the model B-47 bomb bay and high-speed photographing their action subsequent to release. If they hung up in the bay, our point, again, would be proven, this time even more graphically. A net in the exit plane of the tunnel test section would catch the models, so that the downstream propellers driving the tunnel would not be harmed. We made the proposal to the CWT wind tunnel operators, and they agreed to let us proceed, so long as they installed the net.

The next problem we faced was the proper weight distribution design of the scaled model bombs. We either could scale them to simulate the full-scale bomb weight properties, thus assuring an accurate simulation of the bomb's flight path as it moved out of the bomb bay, or we could scale to simulate the dynamic pitching properties of the full-scale bomb. You simply couldn't do both types of scaling on the same model. We argued loud and long until, at last, exasperated, I exhibited, probably for the first time, the leadership ability that was to later make me "famous." I boldly suggested that we make models with both types of scaling, and let the devil take the hindmost! This radical technique of advanced management stood me well for the rest of my technical life, even though I got a lot of snickering behind my back about "coppering my bets." In this case, everyone was happy, since we were only truly interested in bomb motion just after release. Moreover, in the prevailing Sandia atmosphere of financial largesse, expenditure of funds to make a few extra models or $300 per hour wind tunnel runs was never a factor. The AEC had difficulty finding ways to spend the huge sums of money that Senator Clinton Anderson, the patron Saint of AEC activities, and his Congressional cronies, continuously threw at them. For us, this cushy atmosphere made for wonderful, challenging work conditions. We merely asked and hardly ever got a "No!" response.

The two subsequent wind tunnel campaigns completely vindicated my group's theory of what had happened, and also proved out the baffle fix. The static test technique using the balance showed no net up or down forces on the model, just as our theory had predicted. Both types of free flying models hung up in the bomb bay after they were released, just like the real thing. The model baffles, when installed in the model bomb bay, corrected the problem. However, it appeared that due to buffeting, the descending bomb would probably rub against the sides of the baffle, which might cause small yaw oscillations after the bomb cleared the bay. In this event, although the effect on bomb trajectory would be minor, it nevertheless suggested caution.

The baffle fix was never tried. The Air Force instead accelerated the development and installation of the Douglas ejection bomb rack. It was an unqualified success in its first Salton Sea tryout. Later it was a factor, along with Heinemann's designs of the A4D and F4D "Heinemann's Hot Rods"—U.S. Navy high speed fighter aircraft—in Ed Heinemann's winning well deserved national awards as an outstanding and pioneering engineer.

Significantly, our wind tunnel exercise represented the beginning of the serious use of such techniques to study the separation of bombs from aircraft. I wrote and published a long technical paper both describing the technique and proposing more sophisticated computer-driven extensions to it. It turned out to be a landmark paper, and variants of my proposals continue to be used in both wind tunnels and computational simulations.

The Mark 7 on display. The angle irons were welded at the juncture of the tail fins and the conical afterbody. From the National Atomic Museum (Albuquerque).

ANGLE IRON TO THE RESCUE

Warhead design technology in the early '50s had progressed sufficiently to permit the development of a small diameter, externally carried—under the wing or fuselage—atomic bomb compatible with the biggest of the new generation of jet propelled fighter-bombers coming into service. Until then, all strategic A-Bombs were high-yield, large diameter designs that were crammed into the bomb bays of multi-engined Boeing B-50s, and soon-to-be-delivered B-52s or the giant Consolidated-Vultee B-36 bomber aircraft. These new so-called external stores would be carried below the wing or the fuselage of the largest of the forthcoming F 100 series of Air Force—and their comparable Navy fighter-bomber and medium bomber aircraft. Their development was made possible by the ability of the Los Alamos scientists to make ever smaller and more efficient fission warheads. This progress in weight and size also opened the door to the consideration of their use on ballistic missiles, and began a new tactical and strategic weapons race.

Los Alamos had accomplished a major reduction in A-Bomb size by improving on the "squeezing" ability of the high explosive charges that surrounded the spherical mass of radioactive material, thus producing a critical mass density facilitating the requisite chain reaction. The Mark 7 bomb was to be the first of this new breed. Simultaneously, an under-the-wing, under-fuselage, or torpedo bay-carried small diameter, dam buster gun-type A-Bomb, the Mark 8, was being developed by both Sandia and the Navy, at China Lake in North-Central California.

When the experimental version of the Mark 7, the TX-7, began life, my Division was assigned to see it through design, development and test. Douglas Aircraft in El Segundo, an outfit that made conventional, external stores for military aircraft, did the construction. A little over a year later, the first of 3 full-scale development models was ready for an instrumented drop test at Salton Sea. Alas, all did not go smoothly. In two consecutive dummy drop tests, the tails, tail fins and all, broke off at around 8,000 feet altitude. Both failures occurred following release at high altitude (needed so the drop aircraft could escape the atomic blast) from an under-the-belly attachment on the F-101 fighter-bomber. Both test bombs were traveling at very close to Mach 1 when the calamities occurred. We immediately aborted testing the third instrumented dummy unit, until we could figure out what was happening. We had the sinking feeling that the then mysterious and little understood "sonic barrier" had struck again. In those days, anything that happened at high speed that couldn't be readily explained was immediately blamed on the "sonic barrier," just as these days we blame a multitude of mysterious catastrophes on "El Nino."

The ill-fated second test had been scheduled based on the assumption the first failure may have been due to a structural fluke. Following my usual practice in such emergencies, I called key players in my Division (Theoretical Aerodynamics) together to discuss the problem. On review of our design, we could find nothing in our calculations to indicate that we had estimated the airloads on the tail fins incorrectly. Nor could we think of any other reason why the bomb could have been at a higher angle to the wind (in aerodynamic terms, the "angle of attack" relative to the longitudinal center line of the bomb) than the very conservative 3 degrees we designed them for.

Missing from that assemblage was Hal Vaughn, who was on assignment from my Division to the TX-7 program office. On that day, he was on an out-of-town task. His presence, it turned out, would have helped us to immediately understand the source of the problem. Hal, a former ballistician, probably would have known how it was possible for much higher angles to have occurred. But lacking his input, we again relegated another failure to the mysterious "gremlins" that accompanied flight at or near the speed of sound. So began a long and tedious engineering detective process, which eventually resulted in a better understanding of transonic (speeds near the speed of sound) and aeroelastic (the bending of structures under air loading) phenomena.

The TX-7 was a lightweight, thin-skinned design, similar to wingtip and under-wing-carried jettisonable fuel tanks, which were then becoming prevalent on fighter-bombers. Sandia, having little experience with efficient airplane-like construction techniques, decided to contract with an airplane company to do the bulk of the bomb design and prototype/production construction. We selected Douglas Aircraft/El Segundo, specialists in Navy fighter aircraft and, in particular, Ed Heinemann and his Group, to do the job. Ed, as you'll recall, was already famous for his Navy A4D, F4D and other fighter aircraft designs (named the "Heinemann Hot Rods"), as well as being a member of our Aerodynamics Advisory Panel.

When the tails fell off in test, the members of his aero group—headed by the noted Kermit Van Every, and his equally competent Bomb Project Manager, Bob Miller—were as dumbfounded as the members of my group. We again sent out a clarion call for help to our fabulous Advisory Panel to save the day. And once again, the best aeronautical minds in the country gathered at Sandia to inspect the third experimental flight model and advise us what to do now. Along with the original panel members, there gathered new luminaries, each selected for his particular expertise. These included Al Sibila, chief of aerodynamics at Vought (later, Ling, Temco, Vought); Dr. Charles Poor, III, Chief Scientist of the Army's Ballistic Research Laboratory (BRL), in Aberdeen, Maryland; and several other equally distinguished engineers. Following my presentation of the problem, there was much discussion and grilling of me on our design calculations, followed by hand waving and conjecturing. But as it turned out, only two of the experts had a feel for what the problem really was, and it took me a long time and further interviews with one of them to determine the analytical mistakes that were made.

One of the panel heroes was Paul Hill of NACA (now NASA), Langley Research Center in Virginia, who was a leading figure in their PARD (Pilotless Aircraft Research Division), a group dedicated to firing rocket-powered models at transonic speeds. He grabbed one of the 3 tail fins at its tip, and tried to shake the whole dolly-mounted bomb. Indeed, with a vigorous back-and-forth motion, he was able to wag the set-up on the smooth, concrete hangar floor. Paul, ever taciturn and non-voluble, muttered, "It's not strong enough!" Then he sat down and did not elaborate any more. In view of our previously explained design load assumptions, conditions, and detailed structural strength analyses, his statement appeared to have little credence, and the panel more or less ignored his finding. It was not until almost a year later, after a further discussion with both key panelists, that I finally realized the importance and significance of Paul's pronouncement.

The other member of the advisory panel who knew what was happening was a famous ballistician, Dr. Alex Charters. He, too, worked at BRL in Maryland. Alex ran two test ranges, which permitted scale models to go through the speed of sound. They achieved high velocity by firing the models out of cannon barrels. At that time, this was the only way transonic testing could be accomplished, as the first newly invented transonic wind tunnel was still a year away from operation. He said he believed the trouble was caused by a dynamic phenomenon that produced much higher angles of attack than we had designed for, and had nothing to do with transonic speeds, though it would obviously cause the tail section to fail. Since nobody there knew what he was talking about, we also discounted this dynamic interaction process as a possible cause for the failure. As it turned out, both experts were correct, but they were too far ahead of us technically. This was not surprising, since they were the only ones present with ballistics-type experience. It was not until almost a year later, after a further discussion with Dr. Charters, that I finally realized why Paul Hill made his comment, and how Charters' "dynamic phenomenon" figured in.

The other panelists left the meeting still convinced that the old devil transonic shock waves were the source of the trouble, but offered no solution. The results of the meeting left us in a quandary. We had no way to visualize the flow field around the tail section, since Alex Charters' transonic gun ranges, although capable of reaching the desired speed, did not have the precision to accurately determine flight angles of attack of the model bombs as they flew down the range. We also lacked knowledge of a new field called "aero-elasticity," which studied how structures warped under applied air loads. This is the effect that Paul Hill had alluded to, but did not call by its very descriptive name. With much Air Force and Sandia management pressure on us to do something, both Douglas and we agreed on a possible intermediate fix. Jumping on to our interpretation of Paul Hill's conjecture, we decided to beef up the tail structure, even though our calculations showed that its strength was more than adequate for the maximum angles attack we expected could occur during the bomb's flight. The practical fix that Douglas selected was to weld some angle iron between the lower part of the fins and the conical after-body section to which they were attached.

The result was a successful but wobbly drop test of modified dummy bomb unit 3, indicating that the way was now clear to allow the TX-7 to evolve into the Mark 7, as soon as a clean fix to the manufacturing design could be made. We still didn't know how or why the large yaw angles necessary to cause the earlier failures had developed, and I took it on as a personal vendetta to try to get to the bottom of the mystery. It required a trip to Aberdeen, and the talk Alex Charters suggested when he said goodbye to me at the earlier meeting, to finally understand the total picture and thus solve the engineering mystery.

I was a regular visitor to Alex's gun ranges, since they afforded us the only way to test the effectiveness of our anti-oscillation fixes on the big strategic bombs at and above the speed of sound. Dr. "C" ran two ranges: a closed, very long, very low air pressure, 75 millimeter, cannon-fired range; and an open air 155 millimeter howitzer range. In both cases, the instrumentation consisted of flying the models through thin sheets of paper held taut within a frame, and estimating the body angles of attack by analyzing the cutout "cookie-cutter" puncture. For example, if the cutout were symmetrical, the angle of attack would be zero. In the closed range, the gun was aimed towards the ceiling of the tunnel, with maximum range being determined by both the muzzle velocity and effect of gravity on the flying model, since the trajectory was rainbow-shaped. If there were a launch glitch, then the model would scrape the top, negating the test.

In order to provide the correct model weight scaling to simulate the full-scale bomb flight oscillation characteristics, the little models had to be made of a combination of gold and platinum. Each tenth of an ounce of the precious metals had to rigorously be accounted for. As I left Sandia for Aberdeen, each model was carefully weighed on a chemist's scale. When I returned I had to bring back, for accountability weighing, the model remains, smashed flat though they were after impact with the end of the range. Sometimes, due to poor separation from the sabot (a wooden device that grasped the model: was the same caliber as the gun barrel; and broke apart once the combination was out of the gun barrel), the models would fly erratically, and scrape against the wall of the tunnel. In these cases, I had to wriggle down the range with a Polaroid and take pictures of the scratches on the wall. As time went by, I made a wonderful invention to make life easier for me. I fitted the end of the range with enough successive layers of cotton wadding and a mixture of Puffed Wheat and Rice until the models could be recovered almost intact. Some could even be reconfigured and reused. I never did receive the tumultuous recognition due to me for this outstanding and far-out contribution to the art of gun range design, nor did Quaker Oats ever seek me out for an endorsement, or to have my likeness join the late William Penn on their cereal boxes!

On my next trip to BRL, Alex took me aside to explain the TX-7 trouble. The explanation unfolded slowly, and goes like this: The test bombs, the same as the real things, were attached to the drop aircraft via the new Douglas ejection bomb rack. The bomb's topside had two inverted U-shape fittings for the retractable bomb rack "fingers" to engage. The asymmetry caused by these protrusions caused the bomb to take a small flight angle of attack, of about a half degree, and consequently allowed the bomb to develop some lift. This lift, if not cancelled, would cause a trajectory deviation away from the planned zero angle of attack trajectory, and thus cause the bomb to miss the target. To preclude the lift always acting in the same direction, we deliberately made the bomb roll about its long axis by welding small, bendable control tabs on the trailing edge of the fins.

The roll rate that developed led to the dynamic phenomenon that Charters believed caused the angle to amplify beyond the angle for which we had designed the tail section. By pure happenstance, this amplification (called roll-yaw coupling) occurred at the time the bomb, in its fall from the 30,000 feet release altitude, attained the magic sonic speed value. So it turned out the real "devil" was not shock wave phenomena associated with the speed of sound, but rather, the roll-yaw coupling phenomenon. This caused the small flight angle of attack, which resulted from the asymmetry of the U-shape bomb rack attachment lugs, to be greatly amplified—by a factor of 8 or 9 in the TX-7's case—thus exceeding the structural design limit.

Moreover, we had not figured on the other angle of attack amplifying phenomenon that took over as the flight angles increased. This new phenomenon—of aeroelasticity—was just beginning to be understood in the late '40s, as the speeds and consequent air loads took a giant leap upward. Structural warping due to air loading is the concept that applies. In the case of the TX-7, it turned out that the swept-back tail fins, when loaded by increasing angles of attack, warped in such a way as to increase the flight angle of attack over and above the value it would be if they were infinitely stiff. So as the roll-yaw effect caused the flight angle to increase manifold, the aeroelastic elastic effect just exacerbated the situation. What Paul Hill had really meant was that the fin structure was not "stiff" enough!

Our production bomb fix, which turned the TX-7 into the Mark 7, was to beef up the tail structure and change the roll rate, to make the coupling phenomena occur at higher altitudes, where the air loads, because of the lower air density, were less. The United States, busily engrossed in the developing Cold War, now had a high-yield, atomic, external store capability, and the mysterious sound barrier was defeated once again. Our TX-7 mistakes were not repeated later, when we specified the design for the even smaller diameter MK-12 bomb. This new development was done in conjunction with North American Aviation (now Boeing), to be compatible with smaller fighter aircraft, like the F-86 and its successor, the F-100 and the Navy's new fighter aircraft.

Circa 1955, during the TX-12 development, I went with a number of program office people to witness the first atomic test at the new Nevada Proving Grounds. The bomb exploded was a tower mounted Mark 12 warhead. We were in bunkers about 13 miles from ground zero. The anticipated yield was not as great as the Pacific shot I had earlier witnessed, nor did it have the grandeur and immensity of the latter. I do remember the very hot wave that tickled us, simultaneously, it seemed, as the first signs of the explosion. I apparently have suffered no physical distress later in life from these atmospheric tests, at least none of which I am aware, and as I proof read this I am 81.

MICK TAKES A DARE

Commander John ("Mick") Michaelis, USN, the son of an admiral, ran the naval squadron attached to the Air Force's 4925 Strategic Bombing Squadron at Kirtland Air Force Base. It was a small group, equipped with Navy-version F2H2 Banshee fighter-bombers and the Navy's new Douglas A3D twin jet carrier-compatible long range medium bomber. The Banshees assured naval capability to deliver the new under-wing carried atomic bombs, the Mark 12, just coming into the test program. The A3D was specifically designed to carry the new MK 5 bomb in its bomb bay, and was the Navy's answer to the Air Force's Boeing B-47-MK 5 combination.

Mick's squadron worked out of a hangar set aside for the Navy. Like their Air Force counterparts, they conducted bomb drop tests at Sandia's Salton Sea test station, as well as the nearby Los Lunas bomb range, just west and slightly south of Albuquerque. During the course of my work, I made friends with Mick. He knew that I was a WW2 Navy Seaman First Class, and kidded me about having buddies in officer country. Mick was a sandy haired, handsome, young man who exuded a devil-may-care attitude, while simultaneously leaving you with a feeling of his complete competence and professionalism. He may have owned a uniform, but I never saw it. He always had on a flyer's jumpsuit and, with great insouciance, ran an informal ship populated with what I thought to be absolutely top-gun pilots supported by a gung-ho crew. And he is the "can-do" hero of this story.

We were waiting for Mick to taxi up the tarmac by the Navy hangar, all very happy that he had made it back safely from the problematic mission. His crewman helped him down, and he walked over to us with a big grin on his face. "I did it," he yelled,

"Mach 1.1 and nothin' came apart! The old girl bucked and buffeted like a bronco, but the damned probe gave no trouble. I think we got a good one!" We all breathed a sigh of relief and patted Mick heartily on the back and arms. A good day's work! We had a serious problem in detonating the early atomic bombs at the desired altitude over the target area. This test may have provided a possible solution. Here's why:

Paper analyses, supported by the results from prior cleverly designed A-Bomb tests in the Pacific, showed that maximum damage resulting from their detonation would result if the bombs were exploded high over ground zero—anywhere from 2,000-3,000 feet altitude to 10,000 feet, depending on the predicted yield. If they detonated higher, the aircraft that released them might be destroyed by the bomb's blast waves. If they detonated on ground impact, much energy would go into digging a hole, and not destroying potential targets.

A few pounds of TNT properly placed and ignited will easily do in an automobile. A hundred pounds will easily wipe out a typical house; and a thousand pounds, correctly located, will raze a city block. There are 2,000 pounds in a ton. The "yield," or explosive energy of an atomic bomb, is measured in kilotons, or thousands of tons, which is the equivalent of millions of pounds of TNT. The yield of H-Bombs is measured in megatons, or billions of pounds of TNT. The Hiroshima and Nagasaki A-Bombs were "only" of the order of 20 kilotons, but still these "little fellows" wreaked lots of havoc.

The problem for the bomb designers was to devise a way to fuse (i.e., arm the bomb, to preclude premature explosions) and then fire the bombs at the desired "maximum damage" altitude, without resorting to radar altimeters. The latter were taboo, as they could be detected and "jammed" (i.e., given false signals and fired early) by defensive forces. An early solution was to use a timer activated at bomb release. This approach was reasonably feasible, since the delivery aircraft's altitude and speed at release were known, thus establishing the bomb's trajectory and flight time. These now readily accessible, time-of-flight data had been previously determined from drop and wind tunnel tests, and were well documented in the massive computer-generated bombing tables used by the bombardier.

However, all was not that simple. The early bombs, such as the Mark 4, sometimes suffered an unpredictable occurrence of high angle oscillations, which increased the aerodynamic drag, and thus altered the time of flight, causing fairly large dispersions in detonation altitudes. After much study by weather experts and statisticians, it was determined that if we knew something about the weather conditions in the target area, which could be determined by prior reconnaissance, we could correlate atmospheric pressure with altitude. That is, we could model the atmosphere above the target so that each altitude level had a corresponding pressure value. Then, if the falling bomb had the capability of accurately measuring the local atmospheric pressure, it would know its altitude and detonate at the prescribed value. The systems analysts determined that the probable accuracy would result in only 200-300 foot detonation altitude errors—a perfectly acceptable result.

The Theoretical Aerodynamics Group under my supervision had the job to provide the needed atmospheric pressure measurements, through the correct selection of pressure-measuring ports on the bomb's body. We had to find a way to real-time sample and measure the atmospheric pressure as the bomb flew along its trajectory at various speeds, both with and without the annoying oscillations we had not yet learned to tame sufficiently. We spent many hours on wind tunnel testing, recording pressure distributions over the surfaces of the bombs. These were then carefully analyzed at the many speeds and angles of attack that the bomb would have during its drop to the target area. We were able find sampling port positions on the bomb's surface, where the atmospheric pressure was identical to the actual atmospheric pressure at that altitude, with good accuracy, for almost all foreseeable flight situations.

With this new knowledge, we could now arm the bombs when the measured pressure reached a predetermined value. At this signal, a back-up timer, set to detonate a short interval after the predicted firing altitude pressure had been measured, was also started. Thus, the fusing and firing system, further backed up by the original total time-of-flight timer, was practically "fool proof."

However, the ongoing development of heavier, larger yield bombs was causing an increase in their flight speed into the higher transonic range, near the speed of sound. This, in turn, was making it more difficult to measure the true atmospheric pressure with sufficient accuracy. Shock waves, which developed as a result of the near-sonic speeds, were now passing over the pressure measurement ports, causing large changes or pressure spikes in the measured pressure, thus confusing the fusing system. We sought an alternative way to measure the pressure at these higher speeds, which would be relatively unaffected by both the shock waves and the bomb's oscillations.

A professor at Cal/Berkley, E.V. Laitone, had recently proposed using a long, thin, forward-facing probe to be appended to the nose of the bomb, with pressure-measuring ports around its circumference. His calculations indicated that such a probe would suppress the sonic speed pressure spike. We now needed to test it realistically, to see if it would solve our problem. Since the availability of a wind tunnel that could operate through the speed of sound was over a year in the future, we concocted another way to prove out Laitone's idea.

We knew the new fighter aircraft, particularly the Air Force's swept-wing North American Aviation F-86, could attain supersonic speeds during a steep dive from high altitude. We also knew that this was a dangerous, chancy, and little-tested maneuver, which only the very brave would attempt. Nevertheless, the crash dive did represent an easy way both to obtain transonic speeds and test our version of Laitone's probe idea, by dragging it behind the aircraft at the end of a long, thin gage signal-conducting wire.

One evening, while we were both sprinkling water on our adjacent back lawns, I discussed the idea over the fence with my next-door neighbor, Colonel Ritland. Ossie opined that the Air Force and its F-86 fighter-bombers wouldn't touch the stunt with a 10-foot pole, and that he, personally, was against jeopardizing the life of one his pilots, let alone chancing the loss of an aircraft. "We just don't know enough about this 'sonic barrier' thing." He said he would have to go back to Wright Field for permission to make an aircraft modification, and then work with North American Aviation engineers, in order to perform the experiment. This would take forever. And besides, it "simply wasn't in his budget." With a wink, he said, "Bob, why don't you try the Navy? Old Mick might just go for it, and since he is an empire pretty much unto himself, he's not necessarily bound by the same strict protocol that I find myself working under. But," he cautioned, "I'm not sure that those Banshees can hit Mach 1 without falling apart!"

I drove over to the Navy hangar, having made an appointment to see Commander Michaelis, the leader of the resident Navy squadron that supported Sandia in A-Bomb testing. "Mick," I said, "How would you like to do a circus stunt for us?" I told him about my prior discussion with Col. Ritland. I then described the nature of the experiment we hoped he would do. I told him we wanted one of his F2H2 Banshees to drag a probe with a pressure-measuring device, which I would supply, behind it on a thin 200-foot wire line. We then wanted the plane to make

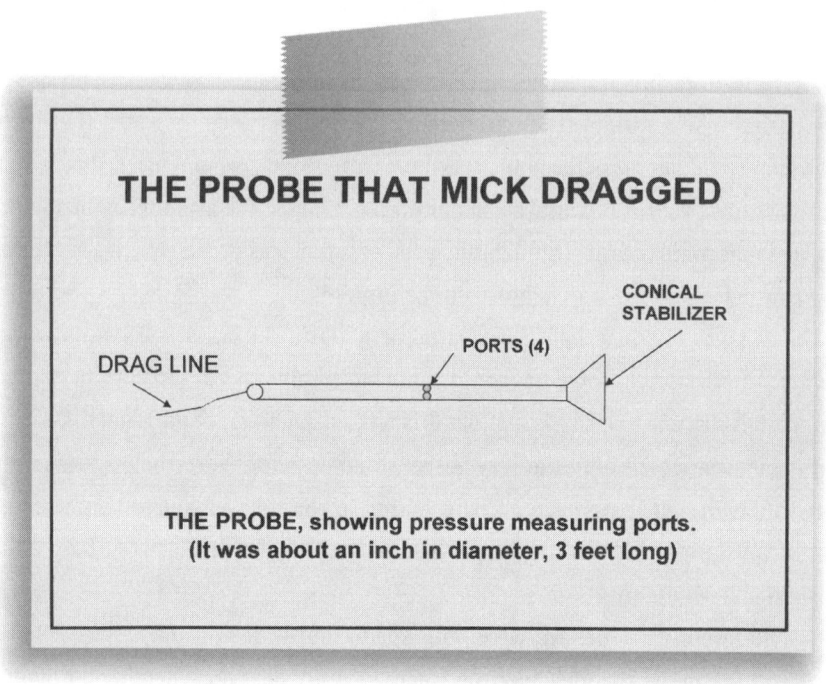

THE PROBE, showing pressure measuring ports. (It was about an inch in diameter, 3 feet long)

a steep dive through Mach 1, the speed of sound, with the probe in tow. We would record the pressure measurements on a tape recorder aboard the aircraft, figure out a way to reel out the probe after take-off, and to dispose of it after the dive.

Mick took this all in. I could sense the "wheels in his head" grinding away. Finally, he smiled mischievously at me and said, "Ski*, that sounds like fun! By God, I'll fly it myself! Talk to my maintenance chief about installation, and we'll schedule a flight over the Las Lunas range." Just like that, it was a fait accompli!

A few weeks later, sporting soul that he was, the crazy bastard made the flight, pushing the F2H2 to its limits. I watched as his modified Banshee rolled down the runway and headed westward. He was back in less than a half hour, emerging from the cockpit with a smile on his boyish face and a "thumbs ups" signal. He later said, "What, me worry? I knew it wasn't going to fall apart!" The probe recorded as planned, and we soon found out that it was a complete success. The sonic-speed pressure spikes resulted in a measured pressure error equivalent to only a few hundred

---

* In the Navy, any name ending in "sky" or "ski" was forever "Ski."

feet of altitude error. We thus had another feasible solution to the critical fusing and firing problem.

Because of his caution and continuing record of good decisions, Ossie Ritland later made Lt. General. I lost track of Mick. He may have made Admiral, but somehow his air of derring-do does not seem compatible with the staid Navy protocol. But he clearly was a leader of men!

LOS ANGELES TIMES

Stellar Awards for "Lord Take Us Through" and "The Country Boy Goes Home" and was nominated for "The Country Boy Goes Home II." Johnson began his career with the record label Nashboro and such hits as "Walk Around Heaven All Day," "He'll Be There" and "Some Days Are Diamonds, Some Days Are Stones." In recent years, Johnson helped bridge the gap between quartets and choirs with the "choirtet." He recorded with the Mississippi Mass Choir such songs as the successful "I'm Yours, Lord." On Wednesday in Tyler, Texas, after a stroke.

■ **Edmund Laitone; Consultant in Aerodynamics**

Edmund Laitone, 85, UC Berkeley expert on aerodynamics and mechanical engineering. Born in San Francisco, Laitone earned a bachelor's degree in mechanical engineering and a master's in applied mathematics from UC Berkeley and a doctorate in applied mechanics from Stanford. From 1939 to 1945, he was an aeronautical engineer with the National Advisory Committee for Aeronautics, performing research at Langley and Moffett air fields. Laitone next went to Cornell Aeronautical Laboratory, working on supersonic aerodynamics. He joined the Berkeley faculty in 1947. In addition to teaching and research, he chaired sections of the engineering school and had visiting professorships in Moscow, Oxford and Xian, China. Laitone remained a respected consultant to such companies as Hughes Aircraft, Douglas Aircraft, the Ramo-Wooldridge Corp., Lockheed Aircraft and General Motors as well as the federal Office of Naval Research. On Dec. 18 in El Cerrito, Calif.

I only met him on one occasion. This notice circa 1995.

THE A-BOMB ENABLES THE SPACE AGE

The *New Mexico Magazine* monthly features an article entitled, "One Of Our Fifty Is Missing," which vividly illustrates that many people still believe the state is either a foreign entity or part of greater Mexico. This contention is fabulously false, and I shall now convince you that the state of New Mexico is so "Yankee Doodle" that it, almost singlehandedly, led the United States into the space age!

In the '40s and early '50s, both the state's finest scientific and engineering minds and its mined raw materials provided the necessities that opened the door to space. The on-going space age is inevitably a child of the atomic bomb, and the atomic bomb is a child of New Mexico. The conclusion is obvious; the space age owes its existence to the state of New Mexico. The chain of logic for this conclusion is simple and relatively straightforward.

The development of smaller-size atomic warheads opened up new vistas in weaponry. Because of their lighter weight, they made possible the accelerated development of airplane-like guided missiles: the short-range Matador, the long-range subsonic Snark, and the long-range supersonic Navajo. My group worked with the companies developing these missiles, to assist in the sensing of atmospheric pressure in their final descent over the target area. Such measurement was required for fusing and firing the warheads. But these slow, vulnerable vehicles were child's play compared to the looming idea of the ballistic missile—a missile that could be fired from a secure bunker, travel large distances and venture into outer space on the way, at very high speeds, and hit an unsuspecting and indefensible target. The ultimate weapon!

The idea of a ballistic missile as a replacement for a long-range artillery shell manifested itself with some reality, in the wake of the V-2 rocket developed by Von Braun and his cohorts in World War II. In truth, the V-2 missile was effective only as a morale buster. It, like the later Mark 4 atomic bomb, couldn't hit the side of a barn door, and even if it did, its puny, high explosive warhead could only inflict minimal damage. It was merely a frightening and very expensive house buster. When measured against its huge fabrication cost, launch facility complexity, and manpower support requirements, it clearly was not a winner.

Nevertheless, the idea of an accurate ballistic missile that could carry a heavier warhead was so intriguing that work on ballistic and other types of guided missiles continued post war, though not with great urgency. The lack of enthusiasm stemmed from obvious bookkeeping analyses. The cost of delivering a relatively small amount of high explosive a long distance away with high accuracy was so staggeringly high as to be unthinkable. More accurately, it was not "cost effective"—the new buzzword for such analyses.

It wasn't until 1953-54, when the scientists and engineers at Los Alamos and Sandia Corporation developed small, relatively lightweight, but high yield, atomic warheads that their accurate delivery by both guided and ballistic missiles became both practicable and cost effective; indeed, they became urgent in view of the onset of the Cold War. At Sandia, a new group of system analysts, heavy with mathematicians, was formed to try to determine the exact ranges and bearings of key USSR targets from contemplated ballistic missile launch bases in the northern midwestern states.

The missile race was now on in earnest, and accelerated Intermediate Range Ballistic Missile (IRBM) and Intercontinental Range Ballistic Missile (ICBM) development began. The next step, the achievement of long duration space flight, became obvious to military planners. Once IRBM and ICBM launch vehicles, compatible with the new warheads, became operational, entry into longtime orbiting space flight was assured. Merely by reducing the weight of the payload sufficiently, the largest of these launch vehicles could readily deliver a payload into low earth orbit. Now the space race was really on!

It is interesting to note that Russia's early atomic warheads were cruder and considerably heavier than the New Mexico designs. These factors forced them to build more powerful launch vehicles for their ballistic missiles. In the ensuing space race, the larger thrust produced by these launch vehicles gave them a marked advantage, in that they could place much larger payloads into earth orbit and towards the Moon. With that extra capability, they inherited, and promptly took advantage of, their edge. They were able to launch Sputnik, and perform other remarkable early space feats, before we could adapt our ballistic missiles and scientific payloads for space delivery.

As the exploration of space continued in the '60s, the two New Mexico scientific laboratories, Los Alamos and Sandia, continued to play a significant role in both military and scientific space missions. In particular, Sandia's contribution to utilization of space is wonderfully documented in a book available from the National Atomic Museum in Albuquerque, entitled "A History of Exceptional Service in the National Interest."

Funny, I was so engrossed in the challenges of producing a new atomic bomb every year that I did not give much thought to space flight and satellites. It was not until early 1957, after I had left Sandia and was participating in the Missile Age, that I again got wind of the new Space Age that was approaching. I stumbled into a meeting in San Diego where grown men were talking about space systems with dead seriousness. By the end of the meeting, I no longer thought they were crazy.

EMPLOYEES' BULLETIN

PUBLISHED MONTHLY BY SPACE-GENERAL CORP., EL MONTE, CALIFORNIA

VOLUME 6, NUMBER 5   MAY, 1966

Twin Aerobee Launches Open New Potential

For the first time, two Aerobee sounding rockets have been launched from the same location in a period of a few minutes.

The twin launching at the White Sands Missile Range, April 14, was made possible by the two towers now in operation there. The rockets were fired exactly eight minutes apart and enabled scientists to conduct a comparative study of upper atmospheric ultraviolet rays.

The second Aerobee tower was moved from Holloman AFB to White Sands last year and has been used several times since October. But this was the first "twin shot" attempted.

With only one tower, the time between launches was more than eight hours.

With dual-launch capabilities, experimenters can now send two separate payloads more than 100 miles high almost simultaneously.

Chief Warrant Officer Lloyd C. Briggs of the Navy facility at White Sands said, "With successful results on our first attempt at a double firing, I expect an increase in the use of both towers simultaneously in the future and expect several more twin shots this year." A Navy crew handles Aerobee launch operations at the range.

The first Aerobee twins reached peak altitudes of 114 and 120 miles.

One contained a Space-General attitude control system operated by personnel of NASA's Goddard Space Flight Center.

Both payloads were recovered using the Space-General parachute system. The experiments were conducted by University of Colorado scientists.

Continued on Page 2

A double exposure photograph captures two Aerobee 150 research rockets in flight. The unique photo was possible because they were launched only minutes apart, rather than hours apart. The "dual launch" of the Space-General rockets from twin towers at White Sands Missile Range was a "first" which enabled scientists to conduct two related experiments.   U. S. Army Photograph

Candidates Visit Space-General

Space-General's 1966 Good Citizenship Drive began this month with the appearance of several California political candidates speaking before employees during the lunch hour.

A Democratic candidate for Secretary of State - William J. Williams was introduced to employees in the Cafeteria, May 24.

Houston Flournoy, Republican running for State Controller, appeared May 27.

Also scheduled to appear prior to the June 7 state primary are:

Jack Solomon, Democrat, seeking nomination as State Senator from the 28th District, June 1.

Continued on Page 3

Simultaneous launch of early sounding rockets at White Sands. From Space-General Employees Bulletin.

A "17" FOR UNCLE JOE

The first hint that something new was up came when Col. Jim Sharp, an Air Force officer who was "on loan" to my group, was put on special assignment by our lab manager. Jim, who had earned his doctorate in physics while a civilian, was a logical choice to supervise the building, in Sandia's huge "south forty" proving grounds area, of a very large dewar jar. A dewar is nothing more than a glorified thermos bottle. This one was to hold a special formulation of liquid hydrogen, which we soon discovered was a necessary ingredient to make an unsophisticated and very large hydrogen bomb. It turned out that the "boys on the hill"—the Los Alamos scientists—felt they could make such a fusion bomb simply by exploding a small A-Bomb in the middle of the liquid. They had scheduled a test in the Pacific to occur about 18 months after Jim's re-assignment. They predicted a yield of 13 megatons—one that exceeded the previous fission bomb high by a factor of 50! Because it was so rudimentary, a disgruntled faction of LASL scientists, led by Edward Teller, the "father" of the H-Bomb, moved to California to concentrate on advanced H-Bomb development. Still under the aegis of U. Cal/Berkeley, as was LASL, they formed the AEC's Livermore Lab, and were supported, engineering-wise, by the new Sandia/Livermore Corporation. Leo Gutierrez, a veteran I had inherited from my previous boss George Hansche, was the first chief engineer of the off-shoot organization. Later, this alliance succeeded in producing practical H-Bombs, by finding suitable metal hydrides to store the hydrogen in a solid form, thus greatly reducing the size and increasing the yield.

In the original planning, the H-Bomb destined to become the Mark 17 would be parachute–stabilized during its fall from the delivery aircraft, in order to allow getaway time and distance for the drop aircraft to escape the heat and shock waves emitted by the explosion. This was an unsatisfactory scenario, as it allowed too much time for both ground and air interdiction possibilities against the slowly descending bomb.

The Mark 17. The later version, after the parachutes were eliminated. From the National Atomic Museum.

Frank Vlasek was the Air Force's premier designer of large parachute systems. He operated out of the then USAF Aeronautical Systems Division at Wright Field, and assisted the Army in developing chute systems to air-deliver big, heavy cargo packages, such as the portable nuclear power plant the Army was developing. He devised a 3-parachute system for the TX-17, a design that was later adapted by NASA to recover the Apollo capsules. However, shortly after his 3-chute system underwent successful drop tests, a Cambridge, Massachusetts outfit, the Allied Research Company, headed by Dan Fink and MIT professor Larry Levy, devised a maneuver that they believed would allow the drop aircraft to successfully live through the H-Bomb's blast, even if the bomb detonation was not delayed by the relatively long parachute drop time. They proposed a 180-degree climbing turn immediately following bomb release, so that the tail of the aircraft was pointing at the expected point of explosion. This maneuver would assure that when the shock wave overtook the bomber, it would not rip off the wings and horizontal tail surfaces. From a military standpoint, a normal bomb delivery was very desirable, and this maneuver, after being proven out in ground tests using high explosive-based simulations, made H-Bomb delivery practicable and essentially indefensible.

Shortly after this favorable string of events, my boss, Ken Erickson, came by and said, "Son, work your old black magic on the TX-17." He said that they wanted to replace parachute delivery with a tail-finned design that would fly right, without the oscillations that hounded the strategic A-Bombs; and they "wanted it right now!" "Easier said than done!" said I, and proceeded to put into gear the testing plans we had been working on, anticipating this call to action.

The monster bomb was to fit into a specially reconstructed single bomb bay in the B-36 heavy bomber. This one-bomb-only bomb bay was created by removing the structure between the two normal identical in-line bays to create, after proper structural reinforcement, the huge, new, single bay, which dictated the allowable length of the bomb. As usual, this alteration left very little span length available for the tail fins needed to stabilize the bomb in its freefall flight. Since reliability dictates precluded the use of extendable or "pop-up" tail fins, we anticipated going into our usual bag of aerodynamic tricks to develop a non-oscillating, or limited oscillation, bomb. I assigned Paul Rowe, one of our top aerodynamicists, as project engineer, to try to find a hurry-up solution to the stability problem before the Pacific test, which was now about 3 months away. I was afraid that, if we failed, a black mark impossible to erase would appear on my until-now unblemished record of making the bombs "fly right."

Our normal testing tool for aerodynamic shape development was the high-speed wind tunnel. We hired either the now-defunct Southern California Cooperative tunnel (the CWT or "Co-op"), on S. Raymond in Pasadena, or the Cornell Aeronautical Lab tunnel, in Buffalo. Both tunnels had 12-foot square test sections and could achieve 95% the speed of sound (i.e., Mach number 0.95), with a 6-inch diameter model bomb installed. This top speed was generally satisfactory, since the present stable of atomic bombs did not exceed Mach 1 during their fall from release altitude. The erratic, oscillatory behavior that we had to fix had heretofore always occurred within tunnel speed limits. But the TX-17, despite being filled with low density liquid hydrogen, was so large that it weighed enough to reach a predicted maximum speed of over Mach 1.1.

Fortunately, a year earlier, John Stack's group at NACA Langley Field, near Newport News, Virginia, had successfully developed the first transonic tunnel installed in the 8-foot test section of a tunnel known as the "Igloo." The proprietor was Axel Mattson. Instead of a solid wall in the working section, Stack provided a slotted wall, with the slots in the direction of the airflow. This permitted both the "swallowing" of the shock waves the model produced and prevented the "choking" of the flow, thus allowing the Mach number to exceed sonic speed, even with our usual 6-inch diameter models installed. Cornell Aero Lab also had devised a 4-foot square perforated wall insert in the throat of their 12-foot tunnel, which also permitted transonic testing of smaller (approximately 2 inches in diameter) models.

Both of these tunnels lacked the ability to take "pictures" of the shock wave patterns at that time. Thus, insight was lacking on studying flow phenomena at sonic speeds with standard optical procedures. We used the same unique test set-up we had used earlier on the A-Bombs to measure the pitching motion of the bomb. Since their bases were flat with large diameters, we could support the models on a swivel, located at their mid-sections, where the center of gravity was located. The swivel joint was attached to a post in the tunnel via a long support arm that went through the opening in the bomb's base. As with the A-Bomb models, we could record the oscillations in pitch after the model was deflected as far as it could go (usually around 5-7 degrees) and instantaneously released. However, without the visual aid, we could not tell the cause of anomalous motions.

At the time of this episode, however, NACA was just bringing on line a second 8-foot transonic tunnel, having a square test section with transparent side walls. The slots were located only at the top and bottom of the working section. This configuration allowed visual realization of the shock wave patterns and tunnel floor/ceiling reflections, if any (the slots were supposed to "swallow up" the model-created shock waves). Using what little span length there was available for our now standard double-wedge tail fins, we intended to try sets of mechanical fixes, which had worked previously, with the slower bombs, in order to control the TX-17's motion.

Since use of both the heavily booked transonic NACA tunnels at Langley Field required calling in serious chits, our test strategy was to first run with our big 6-inch diameter model, in the large test section at the Cornell Lab, up to Mach .92 or .93, and then use a smaller, but still dynamic, model in the transonic insert. The first phase went swimmingly. Paul Rowe excitedly called me to say he had found a combination of fixes that permitted only low angle oscillations throughout the subsonic speed range. However, when he moved the smaller model setup into the transonic insert, disaster struck at Mach number 1.06! Just a little past the speed of sound, the model had flopped to one side of the stop and stayed there—indicating complete instability, at least up to a 5-degree angle of attack. Although we were continually finding out more about flight quirks in the transonic regime, this complete loss of stability was truly unexpected, and nothing in our limited experience could explain it! The model's flight characteristics had been normal all the way from Mach .9 to 1.05! We became suspicious of the small size of the insert. Perhaps there were some shock reflections not swallowed by the porous wall?

Calling on the high priority that AEC programs were always accorded, we gained access, as first industrial occupants of the new 8-foot NACA transonic tunnel, as soon as its shakedown trial "bugs" were ironed out. Paul and I were convinced that our model's action was not "real," but somehow associated with tunnel-generated flow anomalies. The forthcoming crucial test would provide the showdown. Just two days before the scheduled initial H-Bomb warhead demonstration in the Pacific, the greatly heralded tunnel test day arrived. I was in continual phone conversations with Sandia, and told, in no uncertain terms, to be sure to attain flawless performance. As a result, I was in a state of high tension and suffering from fright and dry mouth.

On one side of the transparent side wall of the tunnel was a large viewing area. On this day, it was packed with military, congressional, and AEC representatives, along with the Director of Langley, H.J.E. Reid, his Deputy (and later successor) Floyd Thompson, and other NACA dignitaries. Everyone was aware of what was

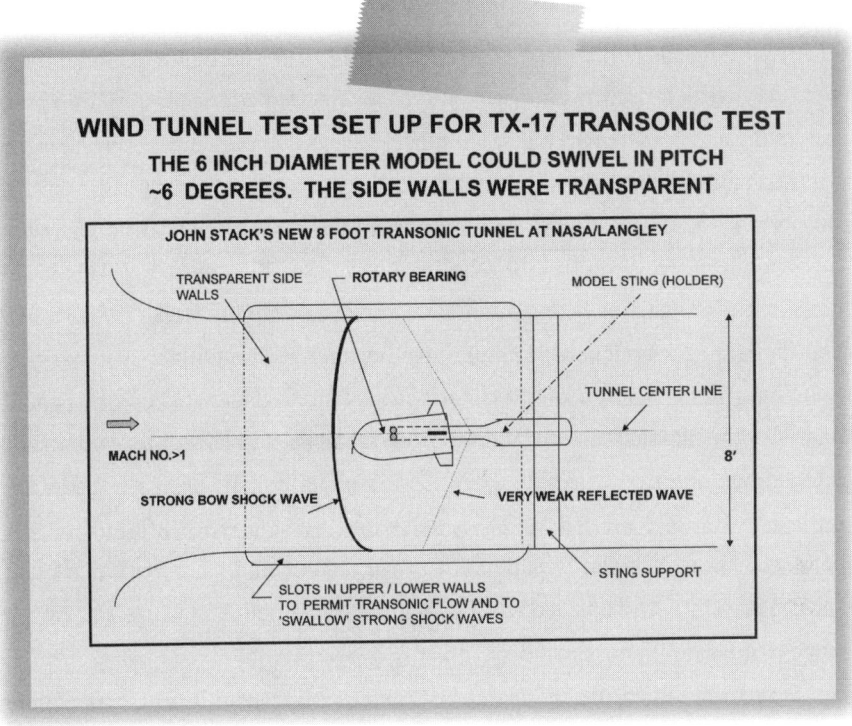

riding on the test. Just before the test was to start, John Stack, the inventor of the transonic tunnel, came up to me and said that the Schlieren system—the optical device that allowed shock waves to be seen on a viewing screen—had temporarily crapped out. Did I want to wait until they fixed it—which could take a couple of hours—or should we proceed? We both anticipated the chagrin of the audience if we postponed. I remained confident in our design and decided to gamble. So, we fired up the tunnel and proceeded up the Mach number scale with fingers and toes crossed, and with eternal hope. At Mach 1.06 all remained well, and I breathed a big sigh of relief. We had done it!

It was a short-lived triumph. At Mach 1.09, without warning, the model flopped over to the stops and stayed there, without recovery, as the speed was further increased to 1.1. The audience got up and walked away. I soon found out that the order to proceed on a parachute-stabilized MK 17 design was given that very day, though with some misgivings. John Stack came over and put a fatherly arm around my shoulders. "Well, Brodsky, back to the old drawing board, eh?" I replied, "No, Mr. Stack, I don't believe it!"

The next day, with the audience gone, Axel got the optical system back on line, we found the culprit that proved Paul and I to be right! Sure enough, there were very weak reflections of the model's shock wave pattern coming from both the ceiling and floor slotted walls. The model had been mounted slightly below the true horizontal centerline of the tunnel. Starting at Mach 1.09, the weak, reflected shock from the floor just hit the trailing edge of the lower set of tail fins, while the corresponding reflection from the ceiling just missed the upper fins. This small pressure imbalance was enough to flop the model over to the stops. We surmised that the same phenomenon also must have occurred at Cornell.

I immediately spread the good news to as many important people as I could find. Alas, the die had been cast. When the test in the Pacific bore out the Los Alamos calculations, the TX-17 became the MK 17, parachute system and all. It was not until later that the more desirable aerodynamic version went into our nuclear bomb stockpile. A lesson in patience was learned but, indeed, a black mark had been entered into the book. And the U.S. had a strong rebuttal to Uncle Joe Stalin's newly acquired A-Bomb capability.

About the same time, my marriage began to fall apart. I had to leave Sandia with my two kids, and regroup.

Opposite: Alan Pope found it hard to believe that Joe Stalin had the A-Bomb. On the other hand, he was a great composer and celebrated Albuquerque's boom town status by describing the wind-blown sand shower we all endured as the bulldozers worked away.

Chapter 3
GUIDED MISSILE DAYS
1956-58

CALIFORNIA, HERE I COME!

The development of ballistic missiles that I was party to at Sandia convinced me that guided missilery was the "next new thing." I decided to get in on the ground floor. The Convair Division of the newly organized General Dynamics Corporation had just become involved with a multi-company U.S. Navy program to develop both offensive and defensive missiles in support of fleet operations. I applied for and got the job. With 6 good years of experience under my belt, my new starting pay was over 3 times my initial offer from Sandia.

The Convair plant in Pomona, California, hired me in the hope that I could solve the control problems their ship-to-air defensive guided missiles were experiencing. They hoped that because of my Sandia-gained reputation of being able to tame unruly bomb peregrinations, I might be able to work some of the same magic on their TERRIER missiles. They had the ill-fate to become unguideable just as they were about to intercept a target aircraft at high altitude. It turned out that I couldn't solve their problem, because their grief turned out to be due to a guidance, not a control, problem. But I did learn the missile business and opened new horizons —such as getting to California, where the aerospace action was blooming.

The missiles being developed at Pomona were shipboard-deployed, and protected both their own ship as well other nearby ships in an armada. I took on the job of Chief of the Aerodynamics Group. My empire consisted of about 12 engineers, 10 "computresses"/technicians, plus two secretaries. For the first time, I had a woman aero engineer working for me—a rarity at that time. We had responsibility for the aerodynamic design and control of the family of missiles.

As was my atomic bomb design group in New Mexico, this bunch was very compatible. We all got along swimmingly. One of the guys, Ed Velton, took a lot of ribbing. He had just bought the first-in-Pomona strange looking "Bug" from Volkswagen of Germany. He argued that it was very saving on gas. We thought he was crazy. It was 1956.

A WONDERFUL EARLY SIMULATION

Convair/Pomona had originated operations at the huge San Diego main plant, and had moved the Guided Missile Division to Pomona in 1955, less than a year before I arrived. They built an engineering/administration building with their own money, but the large manufacturing facilities were leased from the U.S. Navy. The Applied Physics Lab of the Johns Hopkins University, the fount of many of the physicists who founded Sandia Corporation, was the Navy's subcontractor in charge of directing the "Bumblebee" family of ship-defending missiles. They watched over Convair in the development of the original Terrier missile and, now, the Advanced Terrier and the smaller Tartar missiles.

The group dynamics were considerably different in California than in Albuquerque. In New Mexico, all our friends were work friends, mainly because that's all there were then in the northeast sector of the city, near Sandia base, where we all lived. In California, very few of my coworkers lived nearby the plant. Consequently, we found new friends from the worlds of traditional jazz, politics, little theater, horseback riding, sailing and tennis. California offered many outlets.

The '50s saw a burgeoning utility of, and yearly improvement in, digital computers. IBM brought out faster, larger-capacity models right on schedule, one after another; and you had to learn a new machine language with each improvement. At the same time, the earlier computer standard, the analog computers—machines that used a variety of electrical devices, including resistors, capacitors, inductors, potentiometers, and electronic integrators and differentiators, to directly solve mathematical expressions—were also becoming larger and more sophisticated. Those who ran them, and rabidly loved them, were called "Analog Jockeys." Fascinated by what they could do in support of my job, I became a reasonably skilled operator, with the help and instruction of Dr. Dov Abramis. He was a displaced Israeli who ran the computer laboratory facilities, and a teacher par excellence.

I was head of the Aerodynamics Group or, by aerospace tradition, "Chief of Aerodynamics." We were a relatively young outfit, an off-shoot of the Convair airplane plant in San Diego. This was the plant which, at the height of the war, was producing one B-24 Liberator heavy bomber every hour! In Pomona, we developed and manufactured anti-aircraft surface-to-air guided missiles deployed by the Navy, to protect their ships at sea. The first of these, the BW-1 Terrier, had been developed in San Diego, before the plant moved north, and was already deployed in the fleet.

When I arrived on the scene in February 1956, development on the Advanced Terrier was ongoing, and early engineering on another missile, the Tartar, meant for deployment on smaller ships, was just beginning. Like the Advanced Terrier, the Tartar was guided to its target by a homing guidance system. A ship-borne radar tracked the incoming target aircraft, and the radar reflections from the "bogie" were received by the upcoming missile and converted into signals that caused the missile's control fins to move in such a way as to direct it to intercept the target.

The Guidance and Control Group utilized the analog computer to simulate an engagement with a target, or a number of attacking targets. By knowing the responses of the missile to changing control surface movements, and knowing the trajectory history of the missile as it rose through the atmosphere, whose density changed with altitude, the computer could plot the intercept history. It also had to keep track of the target's movement and, by comparing the missile's position with the target's position, could generate the guidance signals that were fed to the 4 control surfaces. Even more wonderful, we could use real guidance and control system hardware, including the actual control fins and their electric motor drivers, in the simulation. We then had a running record of the motion of the control surface tail fins. By adjusting electrical values, called "pot" (for potentiometer, a variable resistance) tweaking, you could check the effect of changing the guidance laws that commanded fin motion.

It was a wonderful simulation, especially because, unlike the early digital computers, which did not have this capability, you could see exactly what was happening right from launch up to intercept or near miss. The flight history of both the missile and its target was plotted on a large table by a high-speed writing pen. More important, when real missile testing began, you could duplicate the analog test scenario, and see if the computer simulation predicted the actual control motions radioed from the flight.

The CEA analog computer—circa 1956. Genuine analog jockey—the master of the domain. Courtesy of Convair/Pomona.

My Aero Group was responsible for supplying two important pieces of information to the simulation. We had to provide an accurate model of the atmosphere, i.e., how the pressure, temperature and air density changed from sea level to above 60,000 feet. We also had to provide the equations of motion that governed the missile flight, specifically how the missile's lift and drag would change with control surface movement. We got the aerodynamic information—the lift, drag, and pitching moment data, as the missile's body and fin angles changed—mostly from high-speed wind tunnel tests. The missiles flew at top speeds greater that 3 times the speed of sound, and the sought-for aerodynamic values had to be determined throughout the range from Mach 0.4 to > 3.0.

We ran the wind tunnel tests at a Naval Ordnance Wind Tunnel in Daingerfield, Texas, next door to the Lone Star Steel Company. The tunnel was operated for the Navy by our parent company. When Lone Star's massive, high-capacity, high-pressure air compressors, which were ordinarily used to feed coal particles into their blast furnaces, were not doing their primary function, their airflow was diverted to the tunnel, to allow continuous testing at supersonic speeds. This was a chancy process at certain times of the year, when torrential rains annually flood East Texas. Even in the good times, there was absolutely nothing to do in Daingerfield, Texas, except wind tunnel test or make steel.

But, as you may have gathered, I fell in love with the analog computer, and learned how to use it. Like the anachronism that I am, I tried to stick with it as the ever increasingly powerful digital machines were gradually supplanting analogs. It was a losing battle. In this day and age, you'll have to look very hard to find an analog computer that is not a museum piece. And, in the next story, you'll see that even the analog could not have foreseen the Tartar's glitch at Mach number 2.

THE GLITCH AT MACH 2

We were developing the Tartar under the watchful eyes of the Applied Physics Lab. Tartar's design was similar to its larger brother, the Advanced Terrier, but it utilized a then unique dual thrust solid rocket motor made by Aerojet. The new motor provided a short, very high thrust burn, followed by a much longer low thrust sustainer burn. Such a scheme obviated the need for a separate rocket booster, which the Advanced Terrier, destined for use on battleships, carriers, and cruisers, had. Because of its smaller size, the Tartar could be deployed on smaller ships, like destroyers.

In both missiles, the purpose of the high-thrust phase of rocket firing was to rapidly accelerate the missiles to a high enough speed to permit effective guidance using its movable tail fins. At high-thrust termination, the missile's lower thrust sustainer motors continued to keep the missile speed increasing, until the sustainer motors also terminated burning. The missile, slowed by air drag, then started decelerating, as it continued toward the target. Of course, intercepts of short- range targets could occur while the sustainer motors were still thrusting.

"Look, fer Christ's sake, the son of a bitch zigged when it should have zagged! What in the world's going on?" Les said to me. I thought a moment and replied, "I don't know, but look—it just did it again!" We were watching a crucial Tartar missile test at the Navy's China Lake test range at Inyokern, California. Dr. Les Cronvich, the savvy head of the Aerodynamics Branch of the U.S. Navy-sponsored Applied Physics Lab, was our customer. I was Chief of Aerodynamics of Convair/Pomona, his contractor. For the first time, the missile was trying to intercept an unmanned WW2 Grumman F4F fighter, converted to a radio-controlled target drone. It missed, veering away just as it was about to hit the target.

The Tartar's new, unique, but heavy, dual thrust motor lowered the missile's lethal range, but it was still a formidable defensive weapon. To regain some of this lost range, the Tartar employed a brand new lighter weight warhead, and this reduction in weight, at small loss in lethality, regained some of the lost range effectiveness.

The Tartar missile and its launcher. The Tartar was about half the length of the higher performance Advanced Terrier. It was controlled by 4 all-movable tail fin (difficult to see in this view). The long white strip that begins slightly forward of amidships is a low span wing which provides maneuvering lift at high angles of attack. From CONVAIRIETY, *the Convair/Pomona employees Newsletter.*

The initial portion of the full-scale flight test program in 1957-8 had proceeded well, with only normal, expected, and easily fixable problems occurring. The first tests did not use an expendable target drone in order to keep the cost of the test program at a reasonable figure. Instead, internally programmed maneuvers were commanded, and the missile's response to the tail fin's movements noted and checked with the simulations. As more flight experience was gained, however, a strange glitch in flight behavior was observed, as the missile accelerated through Mach 2, twice the speed of sound, and later, when it decelerated below Mach 2. In a narrow speed range centered on Mach 2, the missile did not respond correctly to the control surface movements, and, for a short time, seemed to have a mind of its own. In the test we had just witnessed, this glitch on the way up past Mach 2 had actually caused a disastrous "miss." And the impact of this boo-boo quickly reverberated all the way back to the Pentagon.

The guidance and control design experts were quick to disclaim any responsibility for the errant behavior. They rightly pointed out that the missile's guidance system had correctly commanded the proper tail fin motion, but that motion had not produced the expected steering. The problem, naturally, was dumped into my and Les' laps, since the culprit could then be involved only in the aerodynamics. Just as naturally, we were told in no uncertain terms to quickly "find it and fix it!"

The obvious subsequent analytical drill was to carefully re-examine the myriad of wind tunnel data that had accumulated during development, concentrating on the test runs made at Mach 2, with the control surfaces set at various angles. We also reviewed the dynamic analog computer simulations, to look for anything unusual in the suspicious speed range. All this effort was to no avail. The midnight oil was burned, but the mystery remained. We were stumped, both at Convair, in Pomona, and at APL, in Howard County, Maryland.

Some time later, I had occasion to go across the street to the manufacturing facility, with the mission of sprinkling holy water on a slightly warped Advanced Terrier tail fin. On my way to the "accept or junk" decision, I passed by the prototype Tartar production line. Here, for the first time, I was close to the real missile, as opposed to the by now very familiar wind tunnel model. The first thing unusual I noted was a half-inch diameter hole in the missile's after-body skin, about 1-foot forward of, and exactly midway between, the juncture of the missile body and the leading edge of two of the four tail fins. I looked, but there was no such hole on the diametrically opposite side. "What the hell is that for?" I said to myself. I had to ask the same question of several nearby assemblers, until one guy told me he thought it was an exhaust pipe exit. In my mind, I agreed that he was probably right, but that's as far as my thought process got at the time.

It took a few hours before the proverbial light bulb finally lit, and provided the solution to the problem of the glitch. Here's the skinny: The Tartar required a lot of auxiliary power both to run its radar and drive the high-torque, rapid-response electric motors, which drove the tail fins. It required so much power, in fact, that a battery power supply would have been very heavy. The lightweight solution to this problem was to use a high-speed gas turbine to drive an electric generator that provided the necessary power. The turbine was driven by hot gas obtained from burning a propellant. The turbine's exhaust was routed overboard via the mysterious hole that I had just discovered. Obviously, my group had never been shown this design feature, and, as it turned out, the Tartar's detailed designers had not talked with the Advanced Terrier designers about this feature.

What was happening was that the high-speed, narrow, cylindrical stream of hot gas exhausting through the hole was causing a shock wave to form, just as if it had been a solid piece of pipe placed in the supersonic flow. In a small speed range centered about Mach 2, the shock wave produced by this "pipe" intercepted the leading edge of the two nearest tail fins, rendering them ineffective. No matter what fin angles the guidance system called for, the missile could not respond properly. The controllability problem went away as soon as the missile either accelerated or decelerated and the shock wave no longer intercepted the fins. The solution was simple: route the exhaust to the base of the missile, where it would not only act to increase speed, but also reduce drag. When I looked, I found that this is what had been done on the Advanced Terrier.

Is there a moral here? Clearly, yes—and it's one that every analytical design engineer should have driven into his or her psyche right from the crib: Always look at the hardware! This lesson took a while to become an established research and development methodology in my toolbox.

The missile shown on page 78 is the all-movable mid-wing original BW-1 Terrier. It was a beam rider that followed a radar beam illuminating the target. The mid-wing design limited its maneuverability, especially at the final high angle maneuver command which it sometimes could not respond to—resulting in a near miss. It was replaced by the essentially wingless Advanced Terrier and Tartar missiles, which employed homing guidance, thus avoiding the last minute high angle maneuver. They had all-movable tail fin control surfaces. The ill-placed Tartar exhaust port of the story was located midway and slightly forward of two of the tail fins, as better visualized in the picture.

Now, it is established practice to hold what are called "Design Reviews" at critical stages of the R&D process. At such reviews, expert old-timers not involved in the particular program under review learn about the new program. Such a review for the Tartar would have certainly brought out the anomaly of the exhaust port placement.

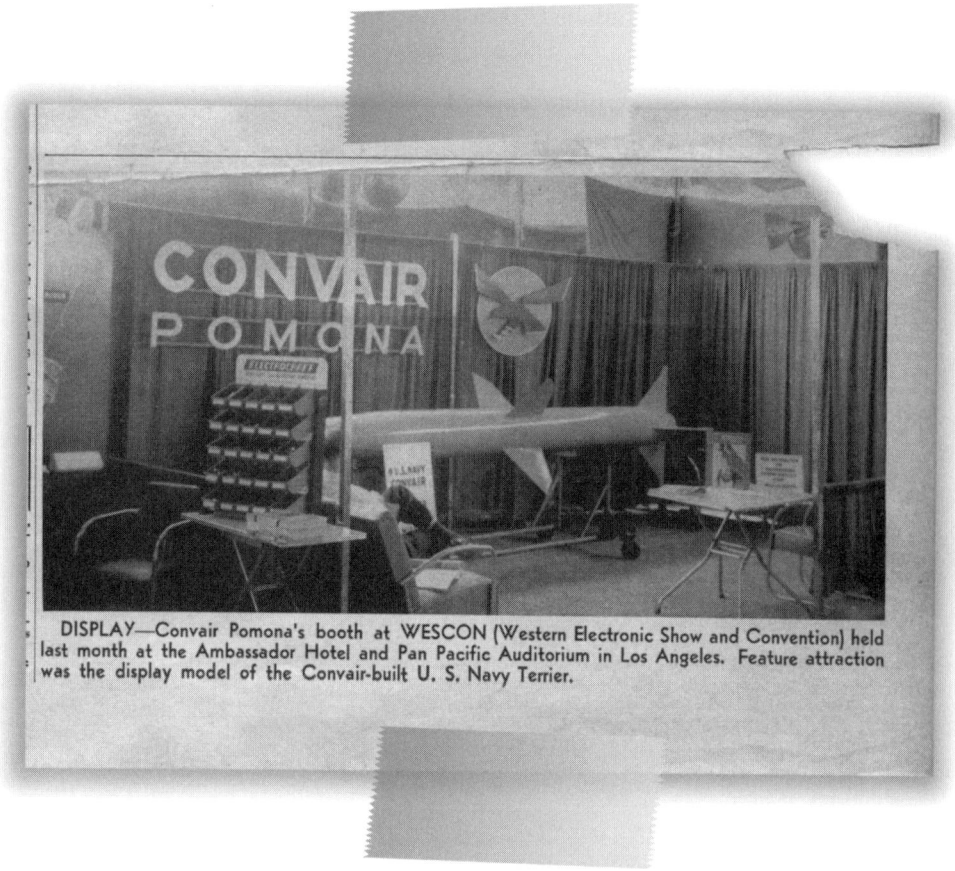

DISPLAY—Convair Pomona's booth at WESCON (Western Electronic Show and Convention) held last month at the Ambassador Hotel and Pan Pacific Auditorium in Los Angeles. Feature attraction was the display model of the Convair-built U. S. Navy Terrier.

Above: *The original (BW-1) Terrier missile. From* CONVAIRIETY, *Convair/Pomona employees newsletter.*

Opposite: *Finding time to get married while studying up on the analog computer and playing boy ingenue Fritz in "I Am a Camera."*

Chapter 4
THE FABULOUS YEARS AT AEROJET
1958-71

THE ROARING '60s!

I left Convair after 3 years, having become a political persona-non-grata with the Division Manager, an avid Republican-minded autocrat, who favored Senator Knowland over Pat Brown for the upcoming gubernatorial race. Anyway, I wanted to go into the just-starting "Space business." I had received and, in late 1958, accepted an offer to become Chief of the Technical Staff in the Space Division of old-line company Aerojet-General, in Azusa, about 10 miles from my Claremont home.

In retrospect, it is hard for me to decide which job yielded the most fun and growth to my career as an engineer, the formative years at Sandia, where everything was new and money was no object, or the beginnings of the space age at Aerojet, where the wonderful laissez-faire attitude of management, coupled with the encouragement of the government procuring agencies, gave their head to anyone willing to take chances and dream big dreams. The '60s were an engineering golden age, culminated by the Moon landings. We may never see another decade like that one!

At Aerojet-General, and then at its newly formed Space-General subsidiary, which I helped found, very imaginative programs on the cutting edge of technology were being undertaken. I was involved in many of them, either directly or as the supervisor, i.e. Chief Engineer, of the troops that both supported the programs and sought my help when they ran into difficulties. We dabbled in everything: The AirTurboRocket, an air breathing engine that turned into a rocket when it left the atmosphere; reverse osmosis, which turned sea water into potable water; the Nerva nuclear rocket engine and the Snap 8 space-borne nuclear power supply; real-time biological agent "sniffers"; instrumentation to be left on the Moon; plus several failures, as well as the unique projects described in this chapter.

For most of the time, I had a wonderful boss, Charlie Roth. He swore like a trooper at the top of his voice, but he knew how to motivate me to work my ass off. Even nicer for me, the higher management of Aerojet/Space-General were friendly and, for the most part, Democrats. Thus, they rooted me on in my political activism, and 10 years later, when a cushy corporate assignment based in Paris arose, I had no lack of champions in high positions.

FLATASS JACK OF JACKASS FLATS

The opener is a story about not seeing the forest for the trees—a lack of foresight that has dogged the careers of many aerospace managers. This is a million dollar example of one such dido, in which I played a key, but not dominant, role. But I clearly share the blame.

In 1960, the new Aerojet-General subsidiary, Space-General, had been formed out of the Space Division in Azusa, where I headed up the technical staff, and a smaller company, Space Electronics, a Glendale outfit that had been formed by very talented Hughes Aircraft alumni, led by Dr. James Fletcher, who would later head up NASA. Dr. Jack Froelich, a key figure in the USA's recent first successful space launch, had come aboard as a Vice President, with a mandate to make the company grow and prosper. At the Jet Propulsion Lab in Pasadena, from whence he came, he had been a very interested observer of the nuclear rocket engine development that Aerojet was performing at Jackass Flats in Nevada.

"Now, let's figure out how to move this mother to Seal Beach with the least excitement. Let's get it down in a report that includes signed and notarized 'Memos Of Understanding' with all affected parties. Let's do it in no more than 6 weeks, so we can fold it in with the total proposal package. Let's do it—Go!" The speaker was Space-General's Director of Operations, the eminent Dr. Jack Froehlich. The problem needing solution was how to transport the assembled Saturn S2 stage—all 30-plus feet in diameter and close to 80 feet in length of it—from the heavy manufacturing plant of Aerojet-General, in Downey, California, to the Pacific Ocean, at Seal Beach, 15 miles away, where it could be loaded on an ocean-going barge for transport to the launch base in Florida.

If we could find a good solution, we stood a good chance at winning what looked to be a very tough competition. The competitors were North American, Douglas, Lockheed, Martin and Convair. The S2 stage was to be the second stage of the giant Saturn V launch vehicle that would send the Apollo astronauts to the Moon.

---

profile..... JACK E. FROEHLICH

Jack E. Froehlich, Vice President and Director, Advanced Design and Experimental Facilities Group, a scientist with numerous firsts in U.S. space and missile programs, directs the group responsible for the concepts in the guided missile and space systems field.

While at the Jet Propulsion Laboratory of Cal Tech, he organized and directed a preliminary study of an earth satellite vehicle for the United States Army's Redstone arsenal. He later headed a group which designed the upper stages of the Redstone test vehicle which was used for reentry experiments with the Jupiter nose cone. Dr. Froehlich also led Jet Propulsion's operations on the Explorer series and on the Pioneer Moon probes.

Before coming to SGC, he served as vice president of the Alpha Corporation, a wholly-owned subsidiary of the Collins Radio Company. Among his duties there were the direction of space tracking and communications, range electronics, range data systems, tropospheric scatter systems and a space communications research programs.

Dr. Froehlich earned his BS and PhD, magna cum laude, at the California Institute of Technology. He has been active in aviation and space endeavors since 1942.

An enthusiastic aviator, Dr. Froehlich enjoys flying -- a carry over from his service as a Marine Corps pilot. His wife, Marion, and his two sons, Mark and John, share this unusual hobby.

Early bio of Flatass Jack. From Space-General News Bulletin.

Froehlich had come to Space-General from the Jet Propulsion Lab, where he had been one of the few leaders, along with Werner Von Braun, who had helped to untarnish the USA's reputation by the hurry-up launching our first successful satellite. This had happened early in 1958, alas, a few weeks after Sputnik, but 3 or so months after the disastrous failure of what should have been America's triumphant conquest of space by the ill-fated Vanguard launch vehicle.

He was a hard driver—acerbic, cocksure, and a bully in an understandable Prussian way. After all, he was a very prominent engineer, and he knew it. We got

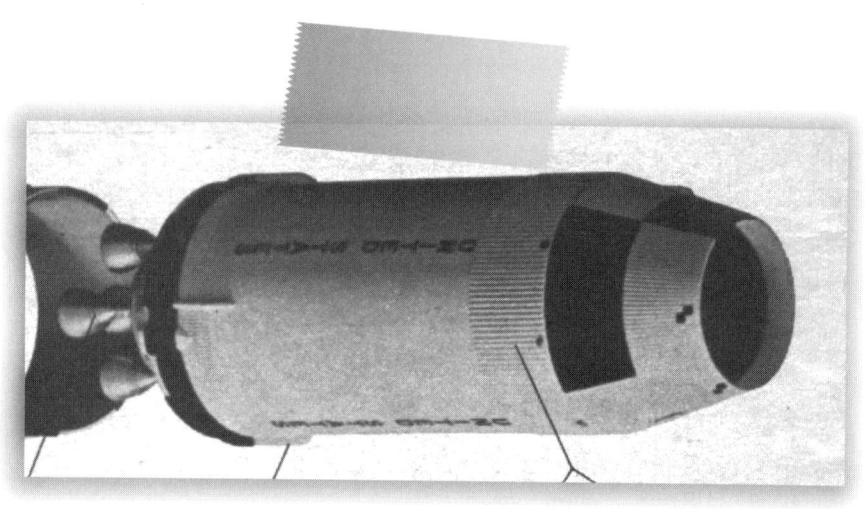

The Saturn V Second Stage, the S2. Courtesy North American Aviation

along wonderfully. I think he liked me because didn't take any guff from him, and always told him what was on my mind. I admired him both for his smarts and decisiveness in a technical world heavily composed of ultra-conservative and slow reacting managers.

His goal was to turn Space-General into a major aerospace industry entity. He was driven to get us more heavily involved in the testing program at the Nevada Test Site, where the Aerojet-developed nuclear rocket engine, the NERVA, was being perfected. And for that, he won the nickname "Flatass Jack of Jackass Flats" from my ebullient immediate boss, Charlie Roth, the VP of Engineering. At that time, my title was Chief Engineer, and I had a number of engineering program managers, test facilities, technical specialist groups, and other service groups reporting to me.

Flatass Jack's opportunity to turn his mandate into reality came along with the anticipated release of the RFP (Request For Proposal) for the S2 stage. For Space-General, by itself, to compete for this billion-dollar program obviously would have been foolhardy. But for Space-General to lead an integrated, all-out Aerojet Corporate effort made some sense. Based on arguments of "right time-right place," Jack was able to convince the corporate powers-that-be to provide the considerable proposal funds and manpower resources that would allow him to lead a successful proposal. It would be, by far, Space-General's most ambitious and most expensive effort; and I, as was my usual role, would he responsible for writing the technical part of the proposal, as well as supplying personnel under me to assist in the multitude of tasks that faced the overall proposal effort.

There was a certain amount of reasonable logic attached to the corporate decision to bid. The other serious S2 competitor was North American/Rockwell, which already was heavily involved in other aspects of the Apollo program. Moreover, of the other possible entrants in the competition, only Lockheed, Martin, and Douglas had launch vehicle, upper-stage experience, but only we had demonstrated engine restart in space, a major S2 requirement, with our AbleStar upper stage.

The favorable bid decision was also made in consideration of some additional advantages. At Aerojet's impressive Liquid Rocket Plant (LRP) in Sacramento, 3 positive factors could help our case: First, the LRP was a serious competitor to Rocketdyne, a North American subsidiary, for the J-2 rocket engines that would power the S2 stage. Since Rocketdyne had already won the F1 first stage engine job, it was again felt that NASA would not want to put too many eggs in one company basket. Second, LRP already had a rocket test facility on the Sacramento River delta, where the complete stage could be test fired—a requirement that we knew would be in the RFP. Finally, Jack argued that a great stock of good will would accrue with the recent return to Sacramento of its Plant Manager, Bob Young. He had just resumed leadership of LRP, cloaked in rave reviews, from a long stint on loan to NASA as chief advisor to their Michoud facility (near New Orleans, on the inland waterway), where they were building the Saturn 1C, precursor to the Apollo launcher, the Saturn V launch vehicle.

One last building block made the whole "megillah" more convincing. In a very foresighted move, Aerojet had recently purchased the Rheem Manufacturing Company, widely known for its home hot water heaters, but actually a builder of very large petrochemical tankage. Its heavy manufacturing plant was in Downey, about 15 miles inland from the Navy base at Seal Beach, from which ocean-going barges would have access to both the Sacramento River delta and, via the Panama Canal, to the Inland water-way, all the way from New Orleans to Cape Kennedy. They clearly had the experience and equipment to construct the over size tankage that would compose most of the S2 stage. The catch, as Froehlich had astutely informed us earlier, was to find a way to transport the completely assembled monster stage from Downey to the sea.

The special transport action team that Flatass Jack organized consisted of some of my engineers, to work on transport fixtures and loading and unloading methodology; marketers and contract specialists, to deal with fiscal and government authorities; and shipping and packaging personnel, to determine optimum routing through the various cities, townships, and fiefdoms that led to the sea.

The engineers determined that that wide-bodied trailers could be readily built to carry the stages and still make cross-street turns, assuming a feasible highway routing could be found. But the along-the-road hazards were all but overwhelming: freeway underpasses were not big enough; telephone and power lines interfered; bridges were not wide enough or had limited load-carrying ability; railroad track crossings were not wide enough. Additionally, a myriad of minor route-particular obstacles all loomed formidable.

About 30 people worked on the routing and memoranda of understanding agreements that had to be made with the Department of Highways, each municipal and county unit involved, the telephone and power companies, various city and state police departments, and the Metropolitan Transit Authority. Several tortuous routing possibilities developed, and were avidly pursued until a complete package could be signed, sealed and delivered. Included were agreements to build special ramps for 90-degree freeway crossings, middle of the night usage of two-lane roads that were converted temporarily into one way streets, removal of some kiosks and bus shelters, re-routing of overhead telephone and power lines, temporary beefing up of a bridge, and rescheduling of some freight train and bus movements.

Deals had to be tentatively cut with city, county, and state government agencies and their local arms, and with two railroad companies. Finally, just making the proposal deadline, we had a sure thing "way to the sea!" Management was tremendously pleased with our effort.

While we were all "asses and elbows" doing our proposal thing and colossally missing the forest, North American was busy lobbying with NASA. They got NASA to agree to their proposal to do the final stage rocket engine firing test at NASA's new first stage engine test facility, along the inland waterway near Bay St. Louis in Mississippi. This not only precluded them from charging NASA for the use of their company-owned rocket engine test site, but also eliminated the problem of moving the stage from North American's assembly site in Long Beach, to the Santa

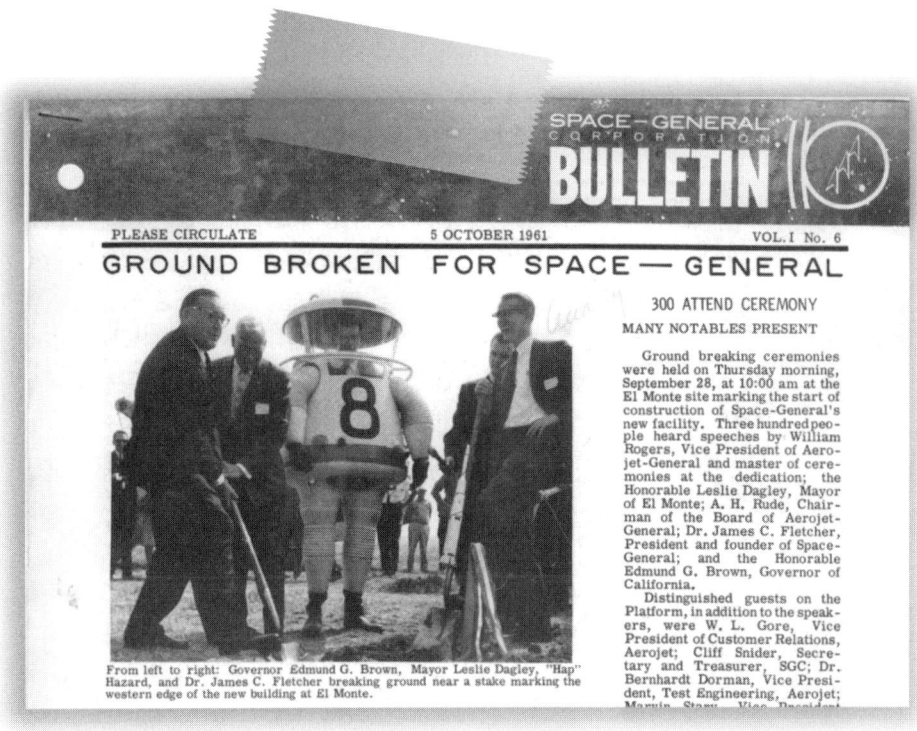

Ground breaking of Space-General, October 1961. Jim Fletcher was CEO. "Hap" Hazard, an old friend from Sandia, is in his space suit.

Susanna Mountain engine test facility, in the San Fernando Valley, which is just as land-locked as Aerojet's Downey plant.

But the crushing down card of their proposal was their offer to build, with company money, a brand new manufacturing/assembly plant on Navy-donated land, right in Seal Beach, hard by the Navy docks, at minimal cost to NASA. Needless to say, they won the competition hands down. When we heard the bad news at the NASA debriefing, Flatass Jack turned to me and said, "Bob, why the hell didn't we think of that?"

A little over a year later, Jack and a son left on a lake fishing/duck hunting wintertime vacation in Idaho, from which they never returned. Only their capsized boat was ever found.

WHO KNEW?

In 1962, space exploration was in its infancy, and Space-General was a very new company. True, we were world class in the development of sounding rockets, like the Aerobees and the Astrobees. But with the exception of Flatass Jack and my new boss, Fred Eimer, both of whom had been involved with the USA's first successful satellite at JPL, none of us had even seen a spacecraft, let alone design and launch one. Thus we stumbled into the new space age as neophytes; and my troops and I had to learn "on the job."

Fred came into my office and said, "Do you think we should propose to make a satellite?" Knowing my technical staff had zero experience in such things, it was easy for me to say, "Hell, No!" Since I already had the reputation for being a bell-weather for important decisions (whatever I opined, you should do the opposite), Fred persisted. He said we had the opportunity to go into the spacecraft business on a sole-source basis through a friendly back door. After I heard the story, I agreed to fire up the technical staff under my command, and together we would learn a new trade.

How this opportunity arose was rather whimsical. In addition to NASA, our other major sounding rocket customer was the Air Force's Cambridge Research Laboratory, north of Boston. CRL was dedicated to research concerning the space environment, in support of the Air Force's fledgling reconnaissance, communication, and spy satellite programs. To this end, they were flying 20 to 30 per year of our Aerobee 150 (100 pound scientific payload to 150 miles into space) sounding rockets, which launched from White Sands, Ft. Churchill on Hudson Bay, Kiruna, Sweden, and other places, and made key scientific measurements during the 5 or so minutes they spent in true space.

In getting into orbiting space flight, Cambridge saw the chance to make such measurements over a very long time—like a year instead of minutes. But they had no charter from the AF to develop or buy spacecraft. Their wily Sounding Rocket Group leader, our longtime customer, decided to make an end run. Since he was authorized to buy sounding rockets and their payloads, he decided to order a new-type sounding rocket—one that stayed in space longer than the normal 5 minutes. As also normal, he gave the order to his usual supplier, so that no one would question the purchase order. This "sounding rocket plus payload" would happen to get into orbit via the low cost Scout launch vehicle, which he also had at his disposal. Thus, the "sporting offer" was made to Space-General to play ball and get Cambridge into space. I'm sure he probably had obtained his superior's tacit, but not written, agreement. Early Space was a little "Wild Westish!"

So we knuckled down to design our first satellite, to be called the OV3. This connoted the Air Force's 3rd Orbiting Vehicle. Of course, we weren't completely ignorant, and F.A. Jack and Fred reached back to the Jet Propulsion Lab, from which they themselves came, to get some excellent spacecraft-experienced, detailed design help. We sought knowledgeable subcontractors to provide the solar cell arrays for our power system, and actually put a tape recorder company into the space supplier business, by helping him get his recorder qualified for operation in the space environment. Temperature control of the satellite proved to be a difficult problem, since so few analytical techniques were available to figure the changes as the satellite went in and out of the sunshine. But we had a large rudimentary passive analog computer, which only one guy could run, that provided some answers. In the end, we proposed to build 4 OV-3's, "guaranteed" to operate a year in space, for the paltry amount of $1,800,000, and subsequently only overran by around 200 Grand. We delivered the first one a little over a year after go-ahead.

As we assembled and tested the 4, some elegant procedures were developed. The best one I remember was the way we balanced them, using a dynamic balance rig similar to the ones that balance auto tires. The OV3 was a spinning satellite, and thus stayed in the proper attitude that it was placed in on spin-up, just as a spinning top does, if it is exactly balanced. Our ace balancers found that a well-placed, well-chewed wad of chewing gum usually did the trick. Once they found the right place to stick the right amount of gum, they sprayed the deposit with a light coat of gunk to assure its permanence. We also found that we could control

Setting up a systems problem on "homemade" Thermal Analyzer are Dave Buchanan and Al Friedman. Resistors and capacitors of various sizes, placed on the board at left, simulate the system under study. The control console in right photo includes amplifiers, variable voltage and current sources, and recording and readout instruments.

Engineers Design Thermal Analyzer As Short Cut to System Solution

The passive thermal analyzer. It consisted of resisters, capacitors and inductors that, when properly connected, can simulate the thermal environment of a satellite in orbit in day and night. Dr. Taghi Mirsipassi was the only one who knew how it worked. From Space-General News Bulletin

temperature, which generally ran towards the hot end of allowable, by temporarily shutting down one or more of the 7 scientific payloads that were being carried.

Each of the OV3s was a great success and all four far exceeded their one-year design goal. Kicking and screaming, I had been dragged into the real space age in the guise of chief engineer of one of America's earliest satellites. None of us really knew what we were doing, but in this case, we were clearly on the right track.

To follow up this act, we next made the OFO, the Orbiting Frog Otolith, satellite. It decisively demonstrated that a frog at least retained a feeling of which way was up in space, even after having undergone a period of aimless tumbling. A frog was used since its ear-related sense of balance system is a close copy (or precursor?) to ours. Later manned space flights, indeed, proved that humans also had this ability. They always seem to know which way is "up"!

SPACE GENERAL TIRE GENERAL — EMPLOYEES' BULLETIN

PUBLISHED MONTHLY BY SPACE-GENERAL CORP., EL MONTE, CALIFORNIA

VOLUME 6 NUMBER 10   NOVEMBER 1966

OV3-2 FLIES EARTH ORBIT
SGC Scores Four for Four With Climatic Last Flight

The Air Force launched Space-General's OV3-2 research satellite on October 28 from Vandenberg AFB, Calif., which, in addition to adding to the Air Force's knowledge of space radiation, will help scientists make a thorough study of the total eclipse of the sun over South America in November.

The OAR satellite, fourth built by SGC for the Office of Aerospace Research, contains five experiments provided by OAR's Air Force Cambridge Research Laboratories (AFCRL) of Bedford, Mass. The experiments are designed to remain active in space for at least a year.

The spacecraft, fabricated by Space-General, is the last of the four satellite OV3 series. All were successfully placed in orbit and all continue to operate.

A Scout rocket carried the OV3-2 aloft following launch from Vandenberg by the 6595th Aerospace Test Wing (AFSC).

Experiments include an electrostatic analyzer which will measure proton and electron spectra; an impedance probe to measure electron density; a retarding potential analyzer to measure electron and ion energy; a plasma probe to analyze positive and negative charged particles, and a mass spectrometer to measure ion species.

Overall purpose of the experiments is to extend existing knowledge of the electron and ion density structure unexpectedly found in the outer radiation belt by the Discoverer XVII space vehicle. The OV3-2 is aiming for an apogee of 780 nautical miles and perigee of 180 nautical miles.

In addition to its primary mission, the satellite will also supplement OAR's study of the South American solar eclipse Nov. 12. OAR scientists from AFCRL

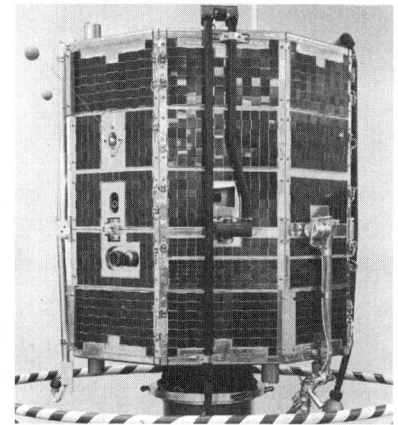

OV3-2 — last of series

are planning the most extensive eclipse study in history in South America to help predict future changes in the aerospace environment which would affect Air Force operations.

The OV3-2 launch was planned so that the satellite would be in orbit during the eclipse, providing data on charged particle variations in the extreme upper atmosphere before, during and after the eclipse. OAR scientists will collect the other eclipse information from aircraft, rockets and ground stations in South America.

AFCRL is technical manager for the OV3-2 construction contract with Space-General.

The OV-3: it was about 3 feet high and two feet across the octagonal. From NASA book, Scientific Satellites.

"FIRST" IN SPACE

No matter what my title and nominal duties, it soon evolved that I became manager and chief writer of proposals for new work. Once I had a streak of 5 consecutive 'wins'—which got me promoted to Chief Engineer—but in no way stopped me from this cutting edge activity. It was a crazy life! You'd work 7 days a week for a few weeks, take a week off, and start on the next one. I loved it, but my family didn't. I never worked on the jobs that we won, leaving this to personnel more experienced in hardware development.

After 5 years in the then- new space trenches, I had my first taste of fame in the February 1, 1963, issue of *Time Magazine*. There, on page 56 of the Science section, was the article that brought me a life-long listing in *Who's Who In America* and my elevation to chief engineer. It covered an invention that was at least 30 years too soon.

There is a clear parallel between an ocean liner adrift in a vast sea and the International Space Station, orbiting alone in space. No ocean liner is allowed to sail without enough lifeboats to accommodate all the people on board; even the Titanic had emergency berths for half its passengers and crew way back in 1912, and that disaster rewrote the maritime laws. So shouldn't the International Space Station, currently under construction, have similar extreme distress facilities?

I thought so more than 43 years ago when my proposal for a space lifeboat made national headlines. I still think so today, as NASA is spending millions on the development of a Crew Exploration Vehicle (CEV), while continuing to ignore the desperate need for a low cost answer to emergency situations.

Shortly after the Space Age began in 1957, Wernher Von Braun predicated the existence of a manned space station in earth orbit by the early '70s. His design looked like a doughnut with a central hub connected to the rim by 3 cylindrical spokes. The entire station rotated about the hub, so that inhabitants in the rim area would feel normal gravity. Such was the enthusiasm for all things "spacial;" and such was Von Braun's tremendous reputation as a seer, that no one doubted the scenario predicted by the great man would happen on schedule.

An artist's rendition, circa 1962, of Von Braun's predicated 1970 space station and the Inflatable Crew Rescue Paraglider. Courtesy Space-General.

In the same time frame, I had recently won a contract to develop an inflatable paraglider, a delta winged vehicle similar to today's sport paragliders, except that the 3 normally rigid booms, to which the wing fabric is attached, were now inflated cylinders. The basic paraglider design had been formulated by Francis Rogallo, so at that time—when paragliding was just becoming faddish—it was called a "Rogallo Wing." Frank worked at the NASA Langley Research Center near Newport News, Virginia, and it was NASA/Langley's idea to send a tightly packaged inflatable paraglider into near space, and have it glide back to Earth. Its two large wing areas were covered with micrometeoroid detectors. There was concern that these small, high velocity particles, known to inhabit near space in showers, might be a serious hazard for astronauts and equipment; so it was important to learn more about their frequency, numbers, and the damage their impact might cause. My proposal for "IMP" (Inflatable Micro-meteoroid Paraglider), to be rocketed into space, whence it would be separated from its delivery vehicle and inflated, was a winner. It would be exposed to micrometeoroid showers for about 5 minutes before gliding back to the ground, where the exposed sensors could be studied.

We soon were busy developing two prototype IMPs, to be launched by my company's workhorse Aerobee sounding rocket. The highest speeds that the paraglider would reach at reentry would be no more than about 5,000 miles per hour, far below the space station's velocity of 17,000 mph. Thus, there would be no re-entry heating problems as occur when re-entering at near-orbital speeds, as a lifeboat would exhibit.

As this development was proceeding, the Air Force's Materials Lab at Wright Field, in Dayton, put out a generalized "Request For Proposal," for the development of new materials and processes to be applied to newly foreseen space age problems. I got the idea that we could alter the IMP design sufficiently to turn it into a space lifeboat, and serve as standard crew rescue equipment for Von Braun's space station. It would be used only in dire emergencies, just as ocean liner lifeboats are employed. It was not meant to be used for the routine return of crew to Earth. I convinced management to make a run at it. When they acceded, I formed a winning proposal team, and came away with a contract that ultimately reached over a million dollars. In those days, that was a lot of money! The tasks that Program Manager Bud Keville and his project team undertook were to design the vehicle; devise and build the specialized tools that would be needed to construct

The full-scale "IMP" being inflated with Nitrogen gas.

it; fabricate one full-scale boom and the nose section; and then subject them both to strength and high temperature tests. The program took 5 years to complete.

We called it "Project FIRST" (for Fabrication of Inflatable Re-entry Structures for Test). I eagerly and proudly told the world about it at a January 1963 technical society meeting, at the Astor Hotel in New York City. Here, a *Time* magazine reporter picked up on it, goaded by my company's always-enterprising PR flack. The article was illustrated by an adaptation of the wonderful artist's proposal rendition (pg. 93) of a partially inflated lifeboat in the vicinity of the space station—a treasure I still have, hidden away in my Fibber Magee-like storage closet.

The proposed space lifeboat concept carried a single passenger, who lay prone in a body canister at the forward end of the central boom. The escapee had little to do, since an autopilot was in control; but passenger over-ride was possible to select a favorable landing site. The whole contraption was stowed in a clam shell-like

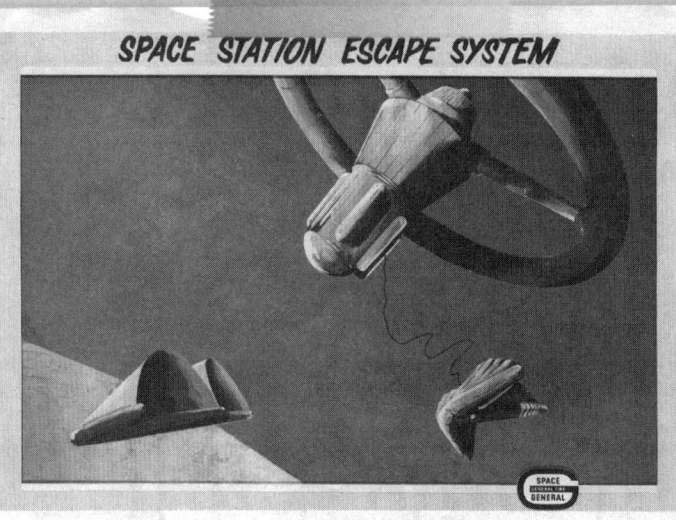

FIGURE 1. Space Station Escape System

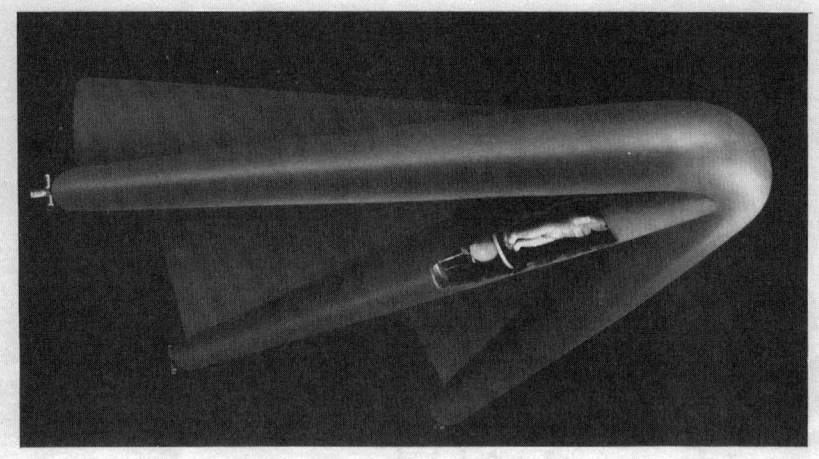

FIGURE 2. Orbital Escape Vehicle Passing Through The Heat Pulse

cylindrical container, whose out-side surfaces were part of the heat protection system for the wing's leading edges. The stowed device weighed about 1,000 pounds, and was 3 feet in diameter and 11 feet long. After release from the space station, FIRST was inflated by nitrogen supplied through a long hose. The 3 wing booms were about 22 feet long and 3 feet in diameter in the nose area.

Technically, the device was a materials wonder—which is why we got the contract. The fabrication process to construct the 3 booms was a marvel of then modern technology. The basic structure consisted of two very pliable bias plies of high strength nickel-chromium alloy meshes. A company that ordinarily made sheer fabrics, like women's hosiery, wove the very thin filament meshes. The metal fabric plies were welded together by a new design high-speed spot-welder, which was able to handle the large hand-fed mesh overlays, in a setup akin to a sewing machine feed.

The boom and nose section were then assembled by fitting the fabric meshes around forms. The "3-fingered" nose section was very intricate in shape, and in some places required 6 plies. Next, the meshes were impregnated and coated with a special liquid silicone formulation from Dow-Corning, which could withstand, by the burning away of the outer layers, the predicted 2,000-degree re-entry temperatures to which they would be exposed. Then, the heat protection material was solidified into the final inflated shapes in special ovens, which could handle the full-scale sections. The specimens were then structurally and vibration tested, first at room temperature, and then exposed to heat lamps simulating the heat distribution anticipated during re-entry. During this development phase, I proposed multi-passenger versions of the basic lifeboat design. The 3-person version weighed about 1,500 pounds. In short, the Materials Lab got its money's worth.

By the time the program was nearing completion, and after several contract renegotiations with the Materials Lab because of program goal enhancements, the climate and mystique surrounding a possible space station had changed radically. A moon landing was within our grasp, and NASA was busy thinking about a follow-on act. The idea of a manned space station was no longer in vogue, having been supplanted by musings about either moon colonization or the germ of the space shuttle idea. The FIRST program gradually shut down, and—humiliation of humiliations—the wonderful spot welder eventually ended up on the scrap heap. But the basic idea did not die!

Thirty years later, after I had retired from a workaday job, NASA began to think that a 3-8 person CRV (Crew Rescue Vehicle) might be a sensible adjunct to their burgeoning International Space Station program. Upon hearing about this, through the auspices of a company that I was consulting for, I made a proposal to revive my earlier lifeboat work that applied modern technology. I later wrote to NASA, via a high-placed friend at the Johnson Space Center, again reminding them of my earlier work. They patted me gently on the head, because their idea of a space lifeboat is radically and very expensively different from mine.

Eschewing the already available tried and true Russian SOYUZ 3-man crew return vehicle, NASA tried reviving '60s technology, and now calling their under-development CRV a Crew Return Vehicle. It weighed 18,000 pounds, and came equipped with hot and cold running mini-skirted flight attendants, and other excessive and sophisticated gear. In various "Letters to the Editor" and to the NASA Director and the Congressional watch dogs, I suggested that while this CRV version may have a rightful place in the space firmament, it is not a lifeboat. I still believe that the latter is really needed, but I appear to be in the minority, since neither NASA nor politicians have reacted to my rantings. Only Bud Keville and I keep the faith.

Alas, my initial flirtation with fame came too late for my father to see, as he had died 6 months prior. It did, however, momentarily alleviate my mother's anguish about my not having become a "real Doctor."

SCIENCE

SPACE
Rescue in Orbit

The first manned space station has yet to be shot aloft, but earthbound engineers are already dealing with the possibility of accident in orbit, struggling with the difficult problem of bringing men back alive from some far-out disaster. What will happen, for example, if a spacecraft's retrorockets are disabled so that it cannot slow down for the long descent toward home? Will the occupants have to be abandoned?

Not necessarily, Aerodynamicist Robert Brodsky told the Institute of the Aerospace Sciences in Manhattan last week. Spaceships, like ocean liners, said Dr. Brodsky, can carry lifeboats. When stowed on board, a Brodsky-designed lifeboat will be a cylinder of strong, heat-resistant plastic up to 1 yd. in diameter, 11 ft. long, and weighing about 1,000 lbs. Inside will be an airtight capsule large enough to hold one man lying face down. A crewman bailing out will crawl into the capsule and detach the lifeboat from the spaceship. As soon as it is clear, nitrogen gas from a pressure vessel will inflate a pair of winglike spars made of heat-resistant woven-wire cloth. As the wings expand, the cylinder will split, forming a heat shield that will protect the leading edges of the wings. The inflated lifeboat will be an air- and space-worthy paraglider (see diagram).

The little spacecraft will have its own retrorocket to start its slanting down toward the atmosphere. Guidance apparatus will do most of the navigation and

PHYSICIST PRESSMAN
Danger in the future distance.

report to the passenger on the success of the perilous maneuver. Dr. Brodsky is confident that the plastic-faced wings can resist the heat of entry into the earth's air. As the paraglider gets deeper into the atmosphere, its speed will drop steadily. At last it will drift slowly near the earth, and the pilot, flying it like an old-fashioned glider, will be able to select a favorable spot on which to land. If he fires his retrorocket at roughly the right time, he may be able to avoid such inhospitable areas as the broad Pacific or the cold wastes of Antarctica.

Contamination Aloft

The possible troubles of space travelers are only the beginning of the space scientists' worries. There is a worrisome chance that moon-bound rockets will contaminate the earth's own atmosphere and cause serious difficulties for the men who stay behind.

The giant rockets designed to boost a man-carrying capsule to the moon will burn more than 2,000 tons of fuel, and a large part of their exhaust gases will be deposited more than 80 miles high, up where the air is only one-billionth as dense as at sea level. Once discharged at that altitude, the gases will not fall for weeks or months, and the air in which they will be floating is so thin that a small amount of contamination can have profound effects. Physicists Jerome Pressman, William Reidy and Winifred Tank of Geophysics Corp. of America have calculated that 25 tons of fluorine can scavenge out of the earth's atmosphere all the free electrons that now make long-distance radio communication possible. Some 25,000 tons of hydrogen, which is soon to be burned in just such vast quantities, could screen off the sun's ultraviolet light, changing the atmosphere's temperature, causing unpredictable and perhaps unpleasant effects on the earth's weather and climate.

ASTROPHYSICS
Primordial Pebbles

Most scientists believe that the solar system—sun, planets and all—condensed out of a vast nebula of gas and dust. Graduate Student Craig M. Merrihue of the University of California at Berkeley is even convinced that some of the first objects that condensed are still around and can be identified. They are "chondrules"—round, pea-sized or even smaller globules of stony material. When they happened to be embedded in meteorites, the tiny pebbles were preserved by the cold and vacuum of interplanetary space and lasted for billions of years.

Merrihue started his research by cadging a 4-lb. chunk of the stony Bruderheim meteorite that fell in Canada in 1960. He crushed it carefully and separated 14 chondrules from the debris. Then he ground the remainder and purified a sample of meteor material until it was free of chondrule fragments. He heated both samples separately and measured the amount of xenon gas that was driven out of them. The chondrules, he found, contained considerably more xenon 129 than the rest of the meteor.

Xenon 129 is a rare xenon isotope that is the descendant of iodine 129, a radioactive form of iodine that was created with the rest of the elements that formed the solar nebula and became extinct not many million years later. Since chondrules contain xenon 129, Merrihue argues that they must have acquired it from the decay of iodine 129. This means that they condensed as droplets during the infancy of the solar system, when everything else in the nebula was dust or gas—and they must be older than the earth or the sun.

PHYSICIST MERRIHUE
Droplets from the deep past.

TIME, FEBRUARY 1, 1963

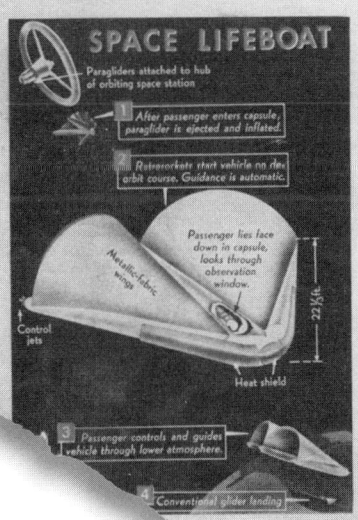

The space lifeboat as shown in Time Magazine.

THE WALKING WHEELCHAIR

In the early '60s, as part of the preparations for the Apollo manned moon landing mission, the Jet Propulsion Lab planned a follow-on program to its very successful hard landing Ranger program, which gave the first close-up views of the moon's surface. The proposed soft lander program was called Surveyor. An initial concept was for it to disgorge a moon surface-traversing vehicle to conduct a neighborhood survey near its landing position. It was to be commanded by a TV–radio control guidance link. This link was to be relayed to and from the moon via the surveyor's earth communication system. A lunar rover competition was opened and the relatively new star on the horizon, Space-General, decided to respond.

"The damned thing looks like a cross between a crazy Bull Terrier and a Peacock!" my boss said. After weeks of in-plant secrecy, we had finally unveiled our robotic entry in the Moon rover sweepstakes. To everyone's amazement and delight, Jack Miller was directing its peregrinations around the back forty of our El Monte campus, using a hand-held radio control link device. With its ridiculous TV camera "head" wagging from side to side on its long tubular swiveling "neck," scanning the local scenery, the 6-legged beast really did look *outré*—like a malformed hound out of the movie Star Wars.

At that time, I was Chief Engineer, reporting to Dr. Fred Eimer, recently recruited from JPL by Jack Froehlich, who, in turn, had reached back there to recruit Jack Miller and Al Morrison for my design group. They quickly had proven themselves to be two absolutely top-notch designers. I organized a proposal team, with the sterling duo in the conceptual design lead. The design we came up with was a walking, rather than a rolling or hopping, vehicle. We believed a walker stood a better chance of navigating an unknown surface than the other options. Its body was triangular in shape—a bit shorter in the forward moving direction than wide—and about 6 inches thick. Its flat sides were parallel to the ground, with the underside standing about a foot off the ground. The swivel-mounted TV camera, which looked like a Cyclops, was mounted on the forward-moving point of the triangle. The aft sets of legs emanated from the other two points of the triangle, thus providing a broad stable base to preclude tipping. More than covering the creature's upper back was a square solar array, which provided drive motor, communications, and TV camera power. This array was hinged near the base of the TV camera supporting "neck," and could be raised up to a 45-degree angle to track the sun as it moved through its lunar day. At its larger angle extensions, it gave the appearance of a sex-deprived peacock! The antenna that communicated with the Surveyor was attached to the top end of the array. Adding to the fowl appearance was a forward thrusting claw at the end of a second tubular bendable "neck," meant to pick up lunar samples, but actually looking more like a peacock pecking at the ground.

However, the feature that made it all work was the unique arrangement and motion of the 3 pairs of biped-like tubular legs, with both knee and hip joints, which gave it the ability to navigate through minor pitfalls such as rocks and potholes. Any larger impediments would be avoided via command from the Earth operator, who would send them in time to transmit a "halt" signal. The motion of each leg set looked like a dog's front legs digging a hole in the ground. The two sets of rear legs were attached to the "haunches" of the triangular body, while the single forward set was attached under the long-necked TV "head." The footpads, themselves, were slightly rounded and serrated on the bottom, to provide a good grip, and thus precluded the need for an ankle joint.

The driving mechanisms that Jack and Al devised were such that there were always 3 of the 6 feet on the ground simultaneously, moving backwards to provide forward motion. The other 3 leg sets, each located right next to and parallel to the foot already on the ground, were making a rapid return cycle, so that they would contact the ground forward, just as the propelling feet reached the end of their backward travel. Steering was achieved by simultaneously speeding up one haunch set of legs while proportionately slowing down the other side. The forward pair of legs maintained the normal pace. If forward motion was impeded before the Earth controllers could act soon enough to stop and regroup, the robot automatically ceased forward travel and waited for a command, perhaps to back up and make a turn before proceeding forward again. We decided to make a full-scale demonstration model of the concept, as part of our proposal submittal. It stood about 4 feet tall from the ball of its foot to the top of its deployed antenna. It was this model that delighted our peers, with its fey look of a defanged futuristic pet. It truly was sensational!

Praised in print by Time, Aviation Week, Life, and newspapers such as the New York Times, SGC's roving lunar vehicle was described as "best of the tribe".

Alas, by the end of the competition, the Surveyor's final design weight had grown so heavy that the intended launch vehicle did not have the lifting power to accomplish the mission. Weight had to be dropped and, to our everlasting grief, the Moon rover portion of the program was summarily deleted. When I heard the news, I almost cried. We had poured so much heart and soul during the frenzied period of the proposal effort. All the participants were simply unwilling to let go of what we perceived as a wondrous device, so we sought ways to revive the program. The beast was kept around the office like a pet.

Months later, the day was saved by the remarkable conversion of the concept from a moon walker to an earth walker—indeed a walking wheel chair that had the ability to go up and down curbs and even flights of stairs! It was Jack Miller, who was acquainted with a researcher at the UCLA Child Amputee Prosthetics Center, who stumbled upon our savior. The UCLA research was sponsored by the well-known Rancho Los Amigos clinic/hospital, which is dedicated to serve such children by developing and testing devices to improve their quality of life. Through discussions and negotiations, Jack convinced them that we could design a walking wheelchair that could not only operate at acceptable speeds on both the flat and normally-encountered slopes, but which also could easily and safely negotiate staircases and confined spaces with simple joy-stick type controls. They were enthusiastic and contracted with us for 3 such walking wheelchairs.

This time the team came up with an 8-legged open-cockpit walker, whose 4 pairs of legs were located where the wheels of a child-carrying sports car's wheels would ordinarily be. It required two battery-driven electric motors, one for each side. The motors drove the leg-pairs through a bicycle-like chain sprocket system. Turns were made by differentially speeding up one side relative to the other, and turns on a dime were possible by reversing leg motion on one side. The whole effect was a natty speedster that begged for the kids to apply a full regalia of car names and stickers worthy of an Indy racer.

The kid drivers reported that they loved them—but, alas, they could not whip down corridors as fast as or do "wheelies" like they could with their conventional electric chairs. One 7-year-old pilot told me, "Yeah, it's great on the steps, but I usually whip around turns at full speed by dragging on the wheel on one side. This tin can just can't do that—it creeps!" To counteract this review, Jack and Al came up with a dual mode vehicle design, which could exchange wheels with legs at the push of a button. But nothing came of this proposal.

A few years ago, a professor at USC, who worked with Rancho Los Amigos, told me that two of the 3 chairs, both bedecked with racing decals, were still operating and were still fun for the kids to drive. But elevators and ramps are now common, and in many cases required by law, thus alleviating the need for curb and staircase climbing by wheelchair dependent individuals.

The walking wheelchair—the kids loved it! Courtesy Space-General

THE SEA BEE SAGA

As the century wound down, there was a continuing, multifaceted competition under way to develop new, inexpensive ways to deliver satellites and other key payloads into space orbit. Over 9 firms in the USA alone were players. Their object was to produce reliable, inexpensive launch systems for small, medium-sized, and large spacecraft delivery. One of the major competitors in the heavy lifter category is the concept of launch-at-sea, a system which, if successful, would permit launch at any desired latitude. But the way this approach is now being pursued may be flawed, as I am reminded by my experience in the mid-'60s with sounding rockets, the precursors to modern day space launch vehicles. Then, we "invented" a sea launch methodology, in a "Keystone Kops" manner, from which lessons can, nevertheless, be learned.

The Aerobee family of sounding rockets were the most successful, most popular, and most used of their ilk, from the first flight in 1946 to the final flight in the mid-'80s. Under various names—Aerobee 100, Aerobee 150, Aerobee Hi, Aerobee 350, and Sea Bee—they made over 1,500 flights into near space. Their function, essentially, was to fly straight up and carry experiments above Earth's sensible atmosphere, where they would make measurements, take pictures, or sample the space environment. The most used of these vehicles, the 150, could carry a 100-pound payload to 150 miles altitude, hence its name "Aerobee 150." These research rocket vehicles also could be equipped with a control system that, for example, could point the experiment in five prearranged directions, either into deep space or towards the ground. The experiments recorded data during the approximately 5 minutes that the payload was in space, where "space" was generally defined as above 60 miles altitude. The 150's payload could be recovered by parachute, and alternatively—to allow reuse of the engine and tankage—the vehicle could be recovered in its entirety. The chief customers for these sounders were the Air Force Cambridge Research Laboratories, outside of Boston, and the NASA Goddard Flight Research Center, outside of Washington, DC. The scientists whose experiments were aboard came from government agencies, universities, and industry, in support of NASA and Air Force research goals.

The Aerobees were developed by my company, the Aerojet-General Corporation, and then by its space research subsidiary, the Space-General Corporation. In the mid-'60s, the Aerobee program was under my purview. Several of the true pioneers of the space business had collaborated on the original Aerobee design concept, and now, at maturity, we were getting orders to produce 50 to 100 vehicles per year. Their flight record was almost perfect.

In order to maintain a near vertical trajectory, the Aerobee had to be launched from a 150-foot tower. A powerful, short-lived rocket booster motor accelerated the 2-stage vehicle up the tower to a sufficient speed to allow its tail fins, like the quills on an arrow, to provide guidance upon exit from the tower. The booster and inter-stage structure then fell away, and the liquid rocket sustainer engine continued to accelerate it, until the propellants were consumed. By this time, the speed was high enough to coast up into space, since, by sustainer burnout, the atmosphere was so thin that aerodynamic drag was negligible.

Two Aerobee launch towers were constructed at the White Sands Proving Grounds, in New Mexico. Others were located at Wallops Island, in Virginia; Fort Churchill, on Hudson Bay, in Canada; at Kiruna, on the top of Sweden; and on the USS Norton Sound, a reformulated cruiser, which operated out of Port Hueneme, near Ventura, California. The extreme Northern launch locations were mostly utilized for the study of the Aurora Borealis phenomena. The towers were expensive to build and maintain. As the program continued, methods to launch Aerobees at other latitudes than the present towers permitted were sought to avoid the tyranny of the limited number of launch sites.

The concept of a launch at sea without a tower had long intrigued some of the founding fathers. Ex-Navy Captain Bob Truax, by then an Aerojet consultant, was a vociferous proponent of this technique, as he remains throughout his lifetime. The idea was to attach a weight to the aft end of the booster motor, on top of which rode the Aerobee sustainer and payload. This combination is akin to the configuration of a standard spar buoy, like many of the usual sea-lane markers. Only the tip of the entire vehicle would stick out of the water. The exits of the booster motor and the sustainer engine nozzles would be plugged, and the plugs then would be blown out on motor and sustainer engine start. As the booster motor accelerated the vehicle, a short launch rail (see figure) enabled the vehicle to remain vertical. The concept was simple, and it did solve a real operational problem. In the mid-'60s, we were able to convince both NASA and the Air Force to each donate one of their already purchased 150's for a sea trial, with support from the Norton Sound. Plans and designs for this radical venture were enthusiastically undertaken by the Aerobee Program Office and supporting engineers.

The finned booster motor attached to the rear of the lash-up. The dimly seen harness allowed the ship's handling equipment to lower it into the ocean. Courtesy Space-General

There was a major difference between the NASA and Air Force Aerobees. The former had 4 stainless steel tail fins, while the Air Force version had 3 slightly larger fins made of magnesium. Nobody thought that the configuration difference would affect sea launch performance characteristics. Nobody remembered, until the "zeroeth" hour, that magnesium and sea water do not live happily together.

On the day of the tests—we planned to fire both vehicles in one day—the Norton Sound steamed out of port a bit late, on what appeared to be a calm summer day. We had to go about 30 miles out to sea to preclude a land reentry impact. Once on site, the grappling hooks on two adjacent ship's cranes were attached to fittings we had welded to the launch rail structure. The cranes gently lifted and then lowered the whole kluge into the water. Only the top 3 feet of the payload fairing, with its stub antenna to receive the fire signal, was above water level. The Norton Sound backed off, and the count down began.

I guess we shouldn't have been surprised that all went magnificently well with the NASA bird, but we were—and elated! In fact, it worked so well that we debated whether it was necessary to launch the Air Force version, especially since it was now late in the day and the sea had risen. The Air Force observer with us made a call back East to his Commanding Officer's home, and then advised us to proceed. The grappling operation was repeated, this time very carefully and slowly, since the ship was now pitching, and 3-4 foot troughs had appeared on the water. By the time we got it into the water and disconnected, it was getting dark and the seas were not conducive to insure a straight-up shot. Reassured by a favorable weather forecast for tomorrow, we decided to cancel, and marked the position with several balloon markers, to make it easy to find next morning. We went back to port and had a few beers, both to celebrate our pioneering success and make a toast to a similar triumph on the morrow.

We found the balloons easily the next day and, as promised, the sea was again flat. Nick Migdal, the Sea Bee Program Manager, asked the Sea Bee divers, who were going to retrieve the marker balloons, to take a look under water, to make sure all was OK. They came back with a strange report. It looked as if there was a lot of boiling going on around the sustainer tail fins! To a man, it hit us—supposed space-age geniuses all—with a truism we had all learned in freshman chemistry. We knocked the sides of our heads as we remembered that water was very corrosive of magnesium- and we had given the magnesium fins a 16-hour bath! "Let's pull it out and take a look at the tail fins," Nick suggested. Again, ship's divers were needed to reattach the grappling hooks, in an operation that was very gingerly completed with success and without damaging the sustainer body. As soon as the cranes had lifted the vehicle out of the water, it became apparent that the fins had suffered drastically. In fact, when we got it aboard for close inspection, they looked like Swiss cheese. We aborted the flight on the spot and called it a day. Of such stuff are heroes made and villains and dolts unmasked!

The denouement of this story is strange. Here we had invented a "better mousetrap"—and nobody came! In retrospect, I suspect that the trouble and expense involved in outfitting ships that could fuel and emplace the rockets was discouraging. We never got a single order for a Sea Bee launch. But today, when an international consortium led by Boeing is spending so much money and betting so much prestige on the success of the huge sea launch platform, which they are operating out of Long Beach harbor, I wonder if they have put their eggs in the wrong basket?

Their floating platform contains a large support complex, replete with a complicated launch tower. Thus, the expense of a land-based launch site is merely duplicated or exceeded, with the only advantage being the ability to launch at desired latitudes. Would not the more simple approach that we devised in the '60s be a better, cheaper answer to achieve the ability to launch at any desired latitude, especially if the magnesium faux pas is remembered?

I wonder, I wonder, I wonder!

DIRTY FRENCH POSTCARDS

It was mid May 1969. I was on my way to my Paris office to take over management of Aerojet's European Operations. I was in my eleventh year of employment, and leaving my position as chief engineer of its subsidiary company, Space-General, which was now being absorbed back into the parent company. To get the appointment to this new position, I had to undergo interviews with the high corporate brass, most of whom knew me. I was able to convince them that my language abilities, admittedly rusty, in French and German would allow me to hold my own, and not be intimidated by the NATO generals and government officials with whom I would be interacting.

With this new assignment came a transfer into the corporate structure, along with the perks that came with an overseas billet—such as housing and home staff allowances, schooling allowances for the kids, auto allowances, and an income tax break. It turned out we could essentially bank all of my salary. "Pretty cushy," I thought!

The reason we had an office in Europe was to support a major NATO contract. It permitted 3 European facilities, in France, Germany and Italy, respectively, to manufacture, under Aerojet licence, rocket motors for their basic aircraft/missile defensive systems. The almost pure profit from each motor built was credited to the Paris office—a beautiful sea of black ink in the otherwise murky financial climate of the times.

The Nov. 25, 1968 edition of the *Aerojet Booster* proudly headlined, "VD Test Device Unveiled," and continued, ". . . called the SeroMatic System, the instrument is a fall-out from biological defense technology developed for the U.S. Army."

The beauty of the device was that it could be operated by a single, relatively unskilled technician, and could test around 800 slides of blood-smear specimens a day. Compare this to the then standard syphilis lab technique of 40 tests per day, which required both a bona fide doctor and an expert technician!

In the mid-'60s, when President Nixon decided the United States would, from this time forward, abstain from the development of biological warfare weaponry, he left my company "holding the bag." We, at Space-General, had been working for several years, under contract from the U.S. Army Biological Laboratories, in Ft. Detrick, Maryland, on a device that continuously sampled the atmosphere, and was capable of detecting traces of either or both of the two popular microorganisms that caused immediate death if ingested over a short period of time. The device worked in real time and quickly rang an alarm that allowed one time to put on the proper gas masks. The time delay between sampling and analysis was sufficiently short to avert death, and the machine we developed was reasonably successful, although it did have an undesirably high false alarm rate.

Were we dissuaded by the "stop work" edict? No, Sir! We turned the machine around and quickly converted it into a peaceful application to solve an age-old problem in the fight against venereal disease.

When I arrived in Paris in 1969 to take over the Aerojet-General Corporate European Office, one of my instructions was to contact the Ministries of Health of all the Western nations, as well as the World Health Organization, in Geneva, with the thought of selling them some machines, and the all-important supply of sera that they required. The latter was a product made from horse's blood. You've never seen a more ridiculous sight than the 50-horse barn and bloodletting lab that had been added on to the El Monte campus of the high-tech rocket-scientist-based Space-General Corporation subsidiary of Aerojet, the outfit from which I came! My instructions said that the machine would sell for $7000 US, and—in very small print—quoted the exorbitant per bottle cost of the precious sera required for the testing.

I immediately noted a strange similarity of reaction at all Health Ministries. When I asked to talk to the person holding the Venereal Disease Desk, I was automatically told that, "Our Country has no VD problem!" I began to feel that I was becoming a purveyor of dirty French post cards. It was only after a bit of cajoling and charming in my smatterings of various languages that I finally gained access to the stolid inevitable VD official. It soon became apparent that, although they were delighted by the existence of a new labor saving device, none of their budgets could stand the capital expense of buying a machine.

At that time, automated medicine machines were in their infancy, with West Germany in the lead. But their automated medicine machines sold for considerably less than our SeroMatic. Thinking about the numbers, I got the idea that we should offer the machines for rent, with the proviso that the lessees buy the sera from us. With this scheme, the rental and sera money could come out of their operating budgets, over which they had considerable latitude. The people back home thought this was a good idea and gave me the go-ahead to deal on that basis. Eventually, in fact, the May, 1970, *Booster* headlined, "Syphilis Test Device Available Free" to qualified medical laboratories! It was clear that the big bucks were in the sera sales.

When I went to the World Health Organization, in Geneva, with this new deal, they quickly glommed on to two machines, one each for their Pasteur Institutes in Paris and Copenhagen. It turned out that syphilis was then rampant in Africa, and they were receiving over 30,000 slides per month to process. They were hopelessly mired down in testing, and were running years late on the analyses. So the machine was as great a success in Europe as it was in the U.S. Aerojet eventually sold the rights, and the business, to the Fischer Scientific Company, which to my present knowledge, continues to produce the machines.

In retrospect, although I accomplished many heroic deeds during my stay in Europe—such as developing a fine taste for French cuisine and wine—I suspect that this may have been my finest hour!

Opposite: *The evolution of an unusual device. By 1970, we were providing the units at no cost to donees who agreed to buy the very expensive sera only from us. All articles courtesy of* the Aerojet Booster

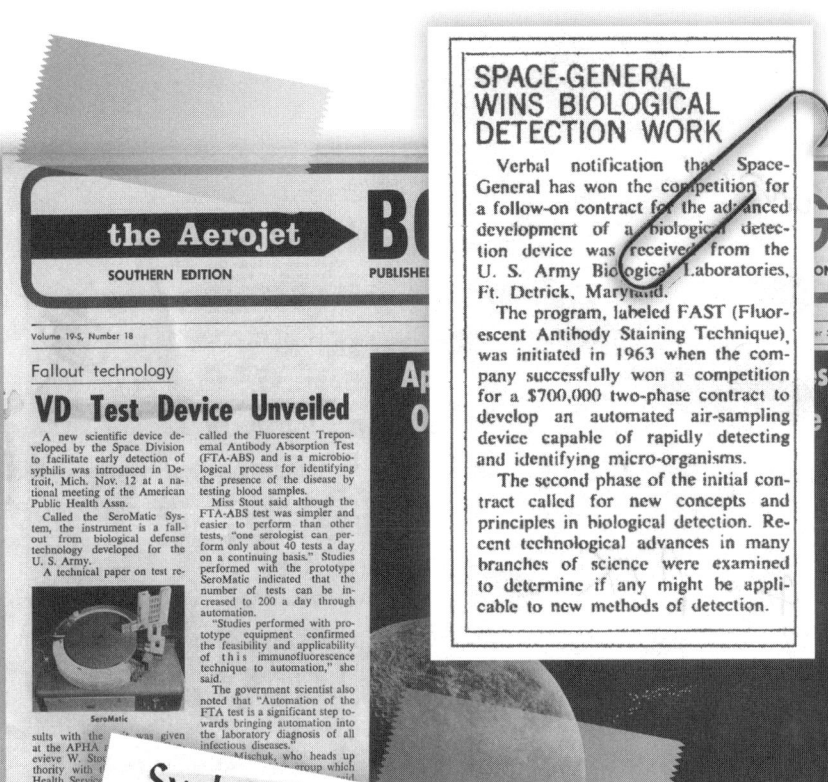

the Aerojet
SOUTHERN EDITION

Volume 19-S, Number 18

Fallout technology

VD Test Device Unveiled

A new scientific device developed by the Space Division to facilitate early detection of syphilis was introduced in Detroit, Mich. Nov. 12 at a national meeting of the American Public Health Assn.

Called the SeroMatic System, the instrument is a fallout from biological defense technology developed for the U. S. Army.

A technical paper on test results with the   was given at the APHA   evieve W. Sto   thority with t   Health Servic   municable   Atlanta, Ga.
The Sero   which looks   cular slide   automates   test for sy

called the Fluorescent Treponemal Antibody Absorption Test (FTA-ABS) and is a microbiological process for identifying the presence of the disease by testing blood samples.

Miss Stout said although the FTA-ABS test was simpler and easier to perform than other tests, "one serologist can perform only about 40 tests a day on a continuing basis." Studies performed with the prototype SeroMatic indicated that the number of tests can be increased to 200 a day through automation.

"Studies performed with prototype equipment confirmed the feasibility and applicability of this immunofluorescence technique to automation," she said.

The government scientist also noted that "Automation of the FTA test is a significant step towards bringing automation into the laboratory diagnosis of all infectious diseases."

Mischuk, who heads up   group which

SeroMatic

SPACE-GENERAL WINS BIOLOGICAL DETECTION WORK

Verbal notification that Space-General has won the competition for a follow-on contract for the advanced development of a biological detection device was received from the U. S. Army Biological Laboratories, Ft. Detrick, Maryland.

The program, labeled FAST (Fluorescent Antibody Staining Technique), was initiated in 1963 when the company successfully won a competition for a $700,000 two-phase contract to develop an automated air-sampling device capable of rapidly detecting and identifying micro-organisms.

The second phase of the initial contract called for new concepts and principles in biological detection. Recent technological advances in many branches of science were examined to determine if any might be applicable to new methods of detection.

Syphilis test device available free

QUALIFIED MEDICAL laboratories can now obtain—free—Aerojet Medical and Biological Systems' advanced automated instrument for syphilis detection.

The instrument, called SeroMatic, employs the FTA-ABS (fluorescent treponemal antibody absorption) test, regarded by health officials as the recognized confirmatory technique for testing blood specimens.

Dr. Eli Mishuck, AMB president, said a cost barrier has been forcing many labs, especially public facilities, to use less effective tests. "Studies involving the most widely used method show that syphilis carriers may be undetected," Mishuck pointed out. "The carriers then continue to suffer personally and unknowingly infect others."

Manual FTA-ABS testing is not in wide use, since it requires eight hours for a highly-skilled technologist to turn out 40 tests manually, an hourly average of five.

Using two SeroMatics, any qualified technologist can turn out 400 a day, or 50 tests per hour.

Speed of the SeroMatic unit cuts lab costs, once the relatively expensive instrument is installed. With the equipment offered free, labs can afford to use the system extensively with a dramatic improvement in test efficiency that would help reduce the destruction caused by syphilis, Mishuck pointed out. The only stipulation is that the receiving labs buy reagents to run the tests from AMB.

The AMB plan calls for free installation and maintenance of the SeroMatic for qualified labs. Training for lab personnel is included, and the instrument would be updated as improvements are introduced. The device will be removed at no charge when no longer needed.

Aerojet Medical and Biological Systems is a newly-formed company located at El Monte, Calif.

THE ALGERIAN GALITZIANER

My office on the Avenue de Neuilly was nicely located near many of the company headquarters of the outfits with whom we dealt. It was also 2 blocks from a key Metro stop, Pont de Neuilly, and only 3 blocks from our home on Rue de General Henrion Bertier. In the grand French tradition, lunch was from 1 To 3 p.m. I strolled home to eat, and my wife and I shared a bottle of wine to ease down the food. We used the car only on the weekends. When I had an office call to make, such as seeing M. Abel Samuel of Sedam once a month, I walked on a nice day or rode the metro on an inclement one. In the mid '60s, Aerojet became involved in the development of air cushioned vehicles—craft that could skim over land or water suspended by a bubble of air. It formed a special division, and won a sizable military contract a few months before I departed for France. In winning they utilized an idea that was owned by Mon. Samuel's company. That's why the the home office directed me to visit with him once a Month, armed with papers that they would send me.

Part of my growing up process was to learn that a "Galitzianer" was a person who had a mean, explosive temper. My parents would use this Yiddish expression over dinner conversations, when referring to various blackguards they had recently encountered. I think the name came from a region of Poland and Austria called Galicia, where such ogres reportedly abounded. During my time in Paris, I met with one such on a regular basis, only he was an Algerian Frenchman; and they, as I later learned, share the same reputation.

Monsieur Abel Samuel, the terrible tempered Algerian, was the President of Sedam, whose headquarters were located within walking distance of my office. Sedam operated a fleet of high speed ACV (Air Cushioned Vehicle) boats, which carried passengers along the Cote D'Azur from Marseilles all the way to Menton, with stops at all the resort towns, such as Cannes, Cap D'Antibes, and Monaco, on the way. The boats were so named because they floated on a cushion of air pumped into a rectangular box-like chamber with no bottom. The air continually escaped around the periphery, so that no part of the vessel came in direct contact with the water. This arrangement caused very little water resistance, enabling the boats to attain high speed with relatively small propulsion engines. Most of the energy needed was required to pump make-up air into the below-deck chamber. Sedam's boats, however, were different in design than the rest of the world's ACVs. Instead of having one, large air chamber, which required a lot of air, they pumped air into a multitude of small diameter tubes distributed around the periphery of the boat. These tubes were called "joupes," after the French word for "skirt." Since they required much less pumped air, their operation was much more efficient.

Before I arrived in Paris to take over Aerojet's European Operations Office, my company had entered into a patent agreement with M. Samuel, which allowed Aerojet to use this novel design in return for a payment of royalties on each unit sold. The agreement covered only civilian applications; military sales were excluded. But shortly after the agreement was signed, Aerojet won a humongous contract from the Marines to build a large quantity of attack ACVs, beating their main competition, Bell Aircraft, mostly by dint of their much more efficient joupe-based design. Their sales of civilian ACVs, based on the Sedam patent, were minimal, as were Sedam's royalties.

When M. Samuel heard the news, he approximated the shrieks of a wounded bull elephant, and his bellows could be heard all the way to California. To placate him, my bosses said they would undertake to revise the agreement, to allow some royalties on military sales, and told him that their soon forthcoming new European manager would be bringing him pages of the proposed new agreement, as they were developed. Thus, the stage was set for the first of my many promised monthly visits to the lair of our now avowed enemy.

The offices of Sedam were near the Arc de Triomphe, and were very plush. A secretary greeted me and showed me into an office, where a very elegant gentleman, who said he was Count Levalle, told me he was the Chef de Bureau and would act as interpreter. I told him I would like to converse in French, even though at this early point in my stay I would have to speak slowly. I showed him the initial few pages of the agreement rewrite that the home office had supplied, which he scanned. He then escorted me into the great man's office, which was also plush.

M. Samuel, a short man with bushy, black hair and penetrating, black eyes, was seated behind a huge desk. He acknowledged my presence and we exchanged pleasantries in the King's French. He then listened passively while M. le Comte Levalle briefed him on the contents of the proposed changes.

Upon hearing the contents of the preamble to the presumed replacement agreement, he began attacking me and my @#$%&* company, with a wrath both enormous and unbelievable. I was so taken back by this onslaught that I reverted to English to object, but was soon imperiously dismissed, with the order to bring back some truly meaningful material on my next scheduled visit. Still shell-shocked, I hobbled back to the office already dreading the next encounter.

After a few months of this same abusive treatment, I went from being an absolute neophyte in choice French cuss words to being a world-class expert, thanks to my office manager, who was well versed in the tongue, and somehow knew what each new blasphemy meant. Despite Levalle's trying to blunt his tirades, M. Samuel continued to rake our fair company and all its executives from stem to stern, ranting that they were stalling, and sending me trash to show him. All our meetings were conducted in French. At first, using French for business was difficult for me, but after a while I became more comfortable, and was able to assuage his rage, by both agreeing with him (for indeed, I did) and making up excuses for our foot sliding.

I arrived somewhat early for our fifth meeting. When I arrived, the office was empty, but M. Samuel's door was slightly ajar. Subconsciously eavesdropping on a phone conversation, I overheard the bastard speaking perfect English. To my credit, I never let him know that I was on to him, and continued struggling on in my version of French, with the Philadelphia-twanged accent that all Parisians hated. A year later, when I was instructed to close down the office and return home, exactly zero progress had been made on the rewrite. I wondered why?

When I was debriefed back in Azusa, my boss implied, but never stated outright, that since we had M. Samuel by the "bondorkas," there was never a serious consideration by the home office of changing the agreement. The stall that M. Samuel correctly perceived was to allow completion of the large order to the Marines. I mulled over this new information with mixed feelings. On one hand, I did not empathize with M. Samuel, who was a veritable living Galitzianer. On the other, I thought that our business ethics might leave something to be desired.

AEROJET BOOSTER

Dynamic 'Z-gram' author
Navy head digs air cushion vehicles

The young, forward-looking top man of the U.S. Navy sees a bright future ahead for air cushion vehicles like those being developed by Aerojet's Surface Effect Ships Division.

Admiral Elmo Zumwalt — who became chief of naval operations at the age of 49 last year — has repeatedly stated that vessels traveling at high speeds on an air cushion show great promise for wide future use by the Navy.

In Seattle recently, he mentioned the 100-ton SES test vehicle Aerojet is building at Tacoma, Wash.

"If these vessels are as successful as we believe they'll be," high-ranking officers, has expressed the belief that such assault craft will play a major role in future naval strategy, since they offer greater tactical flexibility and increase the safety of crews in combat.

New naval techniques that would be made possible by such advanced equipment as air cushion vehicles are needed to counter the threat posed by the growing might of the Russian navy, Zumwalt believes.

The admiral has become widely known as one of the most forward-thinking of all of the list of Chiefs of Naval Operations.

Zumwalt's "Z-gram" directives have made many changes in Navy personnel procedures. Best-known to the public are those which did away with the white hat and bell bottomed trouser uniforms, removed restrictions on long hair and beards, brought beer into enlisted living quarters, and established two-way communication which enables personnel at all levels to make their suggestions —and gripes—known to the top brass.

AEROJET'S amphibious research craft prepares to leave water for land travel.

...assault landing craft, loaded with scale model military

Landing craft for water, land travel

Amphibious Naval landing craft which ride at high speed on a cushion of air—enabling them to zoom over water and then continue over beaches and on inland—have entered the preliminary design stage.

The Naval Air Systems Command has awarded Aerojet a $1.26 million contract calling for preliminary design of such a craft, as well as construction of a full-scale mockup and tests with subscale models.

The craft would be capable of speeds of 50 knots with a nominal payload of 60 tons. While its exact size will be determined by preliminary engineering, it is expected to be about 100 feet long and about 45 feet wide.

Navy officials say the new craft offers an entirely new concept for invasion forces

Dr. R. F. BRODSKY
Manager, European Operations
Aerojet-General Corporation

624.32-50

164, Av. de Neuilly
92 - Neuilly / Seine

THE ROAD TO MAROC

A surprising one-in-a-million event took place on the ferry boat going back to Spain from Morocco. About halfway across the sea, amid all the chatter in French, we heard a loud English-speaking voice proclaim, "My God, it's the Brodskys!" The voice belonged to a good friend, who taught at Pomona College in Claremont, where we normally lived. It turned out he was on a sabbatical at the University of Rabat. What a small world! But first, let me tell you why we were on the ferry.

Dan Kimball, the President of Aerojet-General, was a good personal friend of the late King Hassan of Morocco. Their friendship started when Dan had earlier established a General Tire plant in the kingdom. They were very comfortable and trustful with each other in business dealings. So when Morocco wanted to build a satellite communications ground station to handle all telephone traffic from North Africa, they sole-sourced the job to Aerojet/Space General. But we still had to make a proposal to make it official. I was responsible for the writing of the Business Management volume of this 1968 tome. It was our first job of developing such a ground station.

The station would free the North African continent from the iron grip of the French national telephone company. They were handling all the telephone and wire traffic, via a cable laid on the bottom of the Mediterranean Sea, and had been charging users an arm and a leg for the privilege. Estimates were that the projected satellite communications charges, even after writing off the projected 10 million dollar cost of the ground station, would be about 20% of what they had been paying. The new low cost of communication would make phones affordable to a much larger segment of the North African population. It was clearly a win-win proposition.

After I took over managing the Aerojet office in Paris, I inherited the responsibility of translating the station's operating documents from Space General-generated English into French. I quickly got bogged down trying to do it myself, having particular trouble finding the right French words for the technical components. After a while, I decided to hire a French firm to do the translation. They appeared to be doing a good job, which I closely monitored. After they had produced a couple of hundred pages, I thought I'd better take them down to Rabat. There, I would check their comprehension with the French engineers hired by the Moroccan phone company, both to run the station and train native Moroccan engineers who, eventually, would take over the complete operation.

We—my wife, our 4-year-old son, and I—drove to Algeciras, Spain, took the ferry to Tangiers, and the train to Rabat—a very picturesque trip. Space-General employee Dave Darakjy, who was in charge of station construction, met us at the train station. He checked us into the hotel, showed us around the city, and arranged for shipping, for free, of a newly purchased Moroccan rug, back to California via an Army friend of his who was returning home. I had been in the vicinity of Rabat, at its harbor, Port Lyautey, for a short stay during WWII. However, on my sole liberty venture then, I immediately and violently contracted ptomaine, right after my first dinner ashore, and had to be dragged back to the ship's sick bay; so I had no previous in-depth recollection of the city.

He took us out to the site, about 20 miles NW of Rabat, the next morning, stopping first a makeshift outdoor market, a souk, which was located a few hundred yards from the site. The souk was a sight to behold! It was a throwback to biblical times. Twice a week, the local Bedouin population would ride in on their donkeys, some coming great distances, to shop at the informal market. The Bedouins were handsome, tall people with blue eyes. They "parked" their donkeys in the most raucous, noisy, smelly, lascivious parking lot you've ever seen or sniffed. Two attendants, armed with long boards, tried half-heartedly to maintain the peace, by swatting the males as they attempted to mount the females. Our son couldn't figure out what was going on. The souk had everything a Bedouin could need. There were open fires where you could buy grilled sandwiches, barber shops, pita bakeries, clothing stores, produce and meat stores, wine sellers, etc. The butcher "shop" featured camel's heads, apparently a delicacy. I suspect that their life style had not changed for generations. Dave, who knew their language, told us that the people who frequented the souk had absolutely no idea what the crazy people in the government were building, and they dismissed the construction as nonsense. We took many pictures, using our son as the focus. The Bedouins believed that if you took their pictures, you were robbing their souls.

When we got there, the people at the site were all grousing because the Minister of Communications had not yet provided either power or tap water from the city, as promised. They had to use generators, which were balky, and get water delivered by a truck. This situation existed despite the expected effect of what, I was sure, was a "baksheesh" envelope, which I processed to a Moroccan Ministry through my Paris office once a month, without knowing its contents.

The huge, 90-foot diameter, steerable antenna was being assembled onsite by our subcontractor, Rohr Aircraft of San Diego, using prefabricated parts shipped from the U.S. Moroccan building contractors had constructed the fixed installations. I conducted my business and soon found that the first contingent of French engineers hired to run the station were happy with the translated operations manual portion I reviewed with them. I could, therefore, give a favorable progress report to the people back home, while urging more pressure on the defaulting, but otherwise prosperous, Communications Minister to meet his commitments. Shortly thereafter, both power and water suddenly appeared at the site, but I could swear the monthly envelope was heavier.

That evening, Dave escorted us through the Kasbah. My wife spotted a "tschotke" that she liked, for which the vendor wanted $10. Dave stepped in and said, "Let me do the negotiating." The conversation in Arabic was heated and interminable. Finally Dave said, "Let's get out of here." Sure enough, as we left the shop, the poor merchant came racing after us, agreeing to a $2 sale. Dave was merciless at bargaining—an ability that completely escapes me. He looked on it as a sport. We then ate at a fine restaurant that featured soft couches on which you draped yourself, while watching a wonderful belly dancer, and eating the meat and chicken couscous entrée, by scooping it out of a bowl using your first two fingers. It was delicious.

The station turned out to be a smashing success, and has helped to bring Morocco, Algeria, Libya, Tunisia, and Egypt, all connected by hard line, in better tune with the outside world. The inauguration ceremony, attended by an impressive number of high placed Moroccan and U.S. officials, including the king, was marred only by the fact that the French engineers were still in charge. This situation was corrected a few years later, and the king's 5 million dollar investment in the Aerojet part of the project has been repaid many times.

The Somatelsat Antenna, built by Rohr. The souk was only 200 yards away.
From Space-General Bulletin

CAVEAT EMPTOR

As soon as the NATO Hawk Defense Missile motor production run ended, the black ink that sustained my Paris office books immediately turned blood red. I said to my wife, "The new head man at Aerojet won't notice this for at least a year, so relax!" As usual in things of a cosmic nature, I was wrong! My boss came over, and at a sad meeting at the great seafood restaurant in our Neuilly neighborhood, *Jarasse*, he told us that they had decided to conduct their European business directly from the USA after I closed down the office. We had 6 months to clear out. We would have loved to stay longer in Europe, but I just didn't have enough savvy to find an equivalent, plushy job in such a hurry.

I received no help in finding a job back in Azusa. I made many phone calls to friends, and finally touched base with a former Convair/Pomona buddy, Howard Sommer, who was the newly appointed manager of Aerojet's biggest project: supplying the key remote sensing instrument for the Air Force's Defense Support Program (DSP) satellite. Aerojet and TRW were co-prime contractors on the DSP, the country's most important spacecraft. It provided early warning of the firing of enemy inter-continental ballistic missiles.

The job offered was to oversee the test program for the instrument. I would have preferred a line, rather than a staff, job, but I had little choice, being so far away from the home office.

"We've got a $150,000 bonus for on-time delivery riding on this baby," said Howard, needlessly trying to impress me on the importance of my oversight role in the upcoming test. A hundred and fifty grand was a large amount of "free" money in 1970. The Air Force had insisted on the bonus clause, because our optical detection telescope system was the key component in what was arguably the U.S.'s most important military program, as its more modern successors still are to this day.

We were in the middle of the Cold War, and they were doubly anxious to achieve early deployment of this elegant defensive system. It is now called by the innocuous acronym "DSP," for "Defense Support Program," and is the system of early warning satellites that give the alarm somebody has fired an ICBM—Inter Continental Ballistic Missile—in anger. The detection instrument in question was the first of its breed, and many years had been spent in perfecting its design. Now it was finally ready for its final proof test, and its progenitors at Aerojet believed that we were in the enviable position, vis-a-vis delivery, of sitting fat, dumb, and happy. We had 4 weeks available until we needed to deliver the initial InfraRed telescope instrument to TRW (now Northrop-Grumman Space and Technology), for assembly into their spacecraft for launch. And the only remaining milestone to be accomplished was the forthcoming two week vacuum chamber test.

I had recently returned to Azusa from my assignment of managing the European Operations of the Aerojet-General Corporation in Paris, and was attached to the Program Office. I was in charge of planning the test program, and its successful completion would allow us to meet our contract requirements. The final test of the unit would be an acceptance test, i.e., if the Air Force agreed the test performance met its specifications, then it would "buy" the unit and pay us a bonus, so long as we accomplished the buy-off and delivery before the deadline.

The crucial test was to be performed in the recently completed optical-test vacuum chamber facility of the Orbital Mission Simulator Complex, which the Air Force had funded in support of the program. Our instrument would be placed in the large vacuum chamber, in which the space environment could be reasonably simulated, including the coldness of space and the effect of simulated sunlight as it illuminated the instrument at the various sun angles that would occur in orbit. Satellite eclipse, the situation where the Earth blocked the sunlight, would be simulated by merely turning the "sun" off. The main object of the test was to show that the telescope would always remain sharply focused, throughout the temperature changes that simulated the expected orbital environment.

Our instrument consisted of a long, cylindrical telescope, about 4 feet in diameter, topped by a megaphone-like sun baffle, which was capped by a lid that closed when the sun line approached, looking down the telescope center-line. The "lid" prevented "blinding" the detectors, which could not stand looking directly at the sun. Inside the cylinder, a small, circular reflecting mirror was centrally mounted at the end of 4 approximately 8-foot-long arms attached to its periphery. The other ends of these slender slanting arms were attached to the outside cylindrical wall and main reflecting mirror on its periphery.

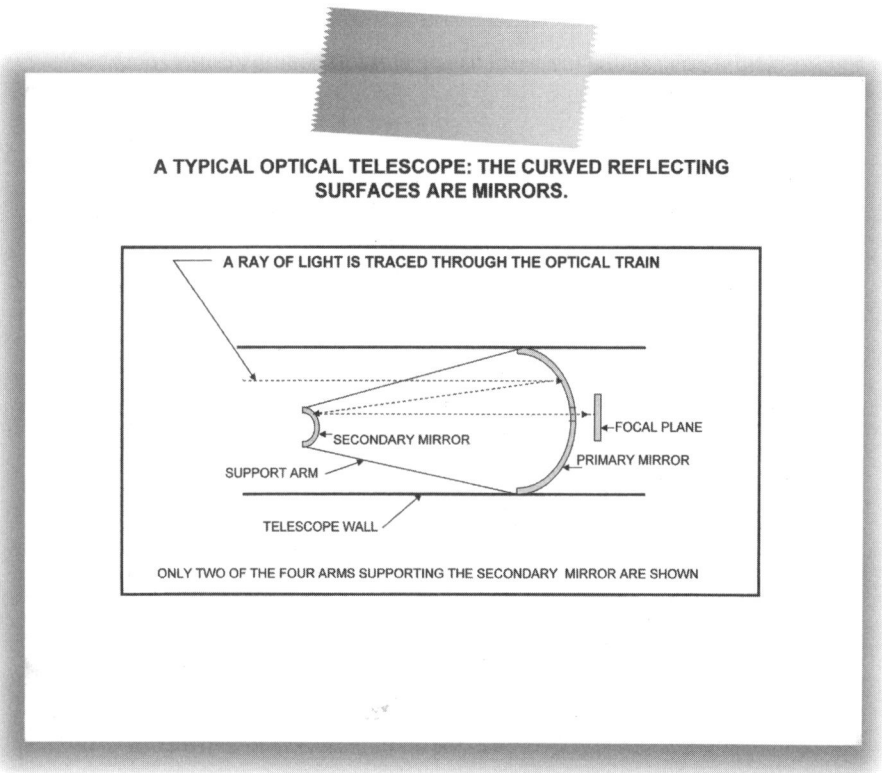

The mirror support assembly had to be very stiff and absolutely unaffected by temperature changes, which would take place as the sun angle changed during the day and when the satellite was eclipsed. If even small changes in the reflector's position occurred, the telescope would go out of focus, and the satellite could not perform its function. This could happen, for example, by the uneven expansion or contraction of any of the 4 mirror support arms.

The hardest problem in the design of the instrument had been to make sure that the focus remained sharp, as the support assembly was alternately either heated by exposure to the sun or reflected radiation from other parts of the spacecraft, and cooled by exposure to the coldness of deep space. Ingenious design solutions, involving the utilization of exotic materials that can compensate for temperature changes, were at the heart of our design.

To minimize the thermally-induced expansion-contraction problem, we selected a special and very expensive steel alloy, called "Covar," as the basic material for the support arms. This alloy, which in its finished form has the same appearance and stiffness as stainless steel, has the added attraction of having, unlike stainless, only a very small change in length as it is heated or cooled. To compensate for even this unallowable small motion, the structures people found a plastic material that shrinks on heating and lengthens with cold; and they used it as part of the arm attachment to the mirror. With this combination—theory, and later, test—we showed that the overall arm length would not change throughout the range of temperatures it would be exposed to in space.

It took a few days to mount the instrument in the vacuum chamber and check out all the test instrumentation. Then, with all being deemed ready, and our still being comfortably ahead of schedule, we started pulling a vacuum. To everyone's relief, we reached maximum vacuum with the telescope still serenely focused on the simulated ICBM rocket exhaust. Next, and the last hurdle in our quest for the pot of gold, was the critical solar simulation portion of the test, wherein the instrument would be illuminated by the "sun" at various angles. Here, disaster struck!

Almost as soon as the "sun" was first turned on, the telescope went out of focus. All hell broke loose and grown men cried! Obviously, there needed to be an immediately designated scapegoat and, by acclamation, the onus fell on the new "French" kid on the block! It was Tuesday, and I was told, in no uncertain terms, that the problem had to be found and fixed by Friday, and that working around the clock was expected.

Suspicion as to the cause of the disaster immediately fell on the small mirror support assembly, and the word went out to concentrate on this possibility first. I called on the technical groups who had designed the instrument to review all their calculations. Several among them had reported to me prior to my overseas assignment, and willingly stepped up to bat. All realized the importance of finding the source of the trouble. We agreed to meet early each morning to discuss progress. By the Thursday morning briefing, it was becoming apparent that the engineers were not going to find a mistake in their design calculations. Gloom filled the room, and I felt particularly distressed. I could vividly see 13 otherwise successful years at Aerojet going down the drain.

But, wonder of wonders, later that same day the problem miraculously went away! The break came from Dr. Max Barsh's Materials Group, which I had established before moving to Europe. I don't know what divination led one of his junior minions to take scrapings from each of the 4 mirror support arms and submit them to the lab for analysis. By Thursday afternoon, they triumphantly reported that only 3 of the 4 support arms were Covar. The culprit 4th arm was made of stainless steel! Without skipping a beat, we removed an arm from the already partially assembled unit #2, and used it to replace the traitor arm. This time, the instrument passed the vacuum chamber test with flying colors, and we delivered with almost a week to spare.

Now started the inevitable formal inquiry into what had happened. It turned out to be a shabby story, but one that made the spacecraft industry seriously, and belatedly, become aware of the joys of both quality control and engineering procedures. In these areas, the ballistic missile industry had pioneered such techniques, which heretofore had been blithely swept under the rug, by the newer satellite purveyors and their government counterparts.

The fabrication history of the support arms started with Aerojet buying, but never checking, what they were assured were billets of Covar from Kaiser Steel, in nearby Rialto. The billets were sent to a forging specialty house that presumably formed them, using Aerojet-supplied dies, into their approximate final dimensions. The arms were then delivered to Aerojet for final machining and assembly into the instrument. During no part of this action was onsite inspection performed or required! It was thus impossible to assign the blame as to how Covar had somehow turned into stainless steel. The guess we made was that somewhere along the line, perhaps even at the source, a Covar arm was damaged and replaced by a stainless one, since the two materials looked identical. All parties interviewed rolled their eyes and innocently said, "Who, me?" Thus are heroes made, paperwork systems greatly and expensively revamped, and source inspection engineers established.

I am happy to report that the junior engineer hero received a handsome bonus, almost on the spot. I, on the other hand, was left with a bad taste in my mouth.

In November 1991, a Defense Support Program (DSP) satellite equipped with Sandia and Los Alamos nuclear detection and environmental sensors prepares to leave the payload bay of STS-44 Atlantis 195 miles above the earth for the transfer to geosynchronous orbit at 22,000 miles.

A later version DSP, ready to launch from the space shuttle. The Aerojet-supplied infra-red telescope instrument and its megaphone shaped sun-shade rides on top of the TRW-supplied spacecraft. Courtesy of Sandia National Laboratories.

Chapter 5
FREEZING IN IOWA
1971-80

DOWN ON THE FARM

Shortly after the DSP fiasco, recounted in the prior story, I got a call from an old Sandia friend, Paul Rowe, who was now working at Aerojet in Sacramento. He told me that his Alma Mater, Iowa State University, had put out a call to its alumni to help in the search for someone from industry to become a professor and head of their Aerospace Engineering Department. Would I be interested?

I had been an instructor at New York University, and in the early '60s had taught graduate courses for the UCLA Engineering Extension organization. I enjoyed teaching and liked the idea of becoming a "real" professor. But, Iowa? Right "after they'd seen Paree?" Well, it seemed like such a long shot that the family said, "What the hell, give it a go." We all realized the chances I would get the appointment were small. Surprisingly, after a visit from the ISU Dean of Engineering at our home in Claremont, we were invited to Ames for an interview. Even more surprising, I got the job!

The move to Iowa was quite a culture shock, both socially and career-wise. About the first, we had the initial feeling that, in one fell swoop, our arrival had doubled both the Jewish and Democratic population of the state of Iowa. In time, we found there were no discernible prejudices in the state, where minorities of all varieties were so small in number they easily were welcomed into the community. Politically, we quickly allied ourselves with Sen. Harold Hughes and his equally Democratic successor, John Culver, and helped Sen. Tom Harkin win his first seat. But academia was the antithesis of industry. Every faculty member was a star (or so they thought); and even though I was undisputed, the department Head for life (as opposed to department Chair, a 2-5 year renewable appointment), the only real hold I had over my faculty was in the annual doling out of the meager salary increases.

My outside activities consisted of joining Rotary; being active in the symphony orchestra festival events; being active in state politics; sailing in the 16-foot "Demon" sailboat that we bought in Kansas City on our way to our new home; and inaugurating my illustrated New Orleans Jazz lecture series, "Serenades for Mouldy Figges." My wife also broke new ground. She served hard liquor at the traditional annual year-opening Department party. We loved Iowa and the people, but hated the winters.

THE ICEBERG COMETH

In mid-August 1999, an iceberg the size of Rhode Island was reported to be a shipping hazard in the near-Antarctic sea lanes. It definitely was not one of those little fellows, the like of which sank the Titanic. This baby, at more than 1,200 square miles, looming 500 feet above the water surface, was a behemoth! It had broken off from the Ross Sea Antarctic Ice Shelf, and was leisurely making its way to warmer waters. The resulting media frenzy reminded me of events that took place in landlocked Ames, Iowa, more than two decades earlier.

It was billed as the "First International Iceberg Utilization Conference," and was incongruously held at Iowa State University in October 1977. At least two people must remember this historic meeting vividly. One was HRH Prince Mohamad Al-Faisal, scion of the Saudi royal family, for, indeed, he was its instigator and chief backer. The second was me, of course, who—as the local expert in matters of remote sensing systems—headed up this area of Conference inquiry.

Saudi Arabia is a country that has more money and oil than water. Its population and growth is severely limited by its inadequate water supply. Prince Al-Faisal was then responsible for his country's water resources, the bulk of which were provided by nuclear-powered boilers that distill sea water. This process yields a sure, but limited and very expensive, supply. A few years earlier, U.S.-educated Al-Faisal had read the early '50s speculations of John Isaacs, of the Scripps Institute of Oceanography, about the feasibility of towing huge icebergs from Antarctica to arid countries. Once anchored offshore, they would provide a bounteous supply of melted fresh water, siphoned from their exposed top surface.

Intrigued by the possibility that this scheme might solve Saudi's problem, the

Prince asked his friend and consultant, Dr. A. A. Husseiny, Professor of Chemical and Nuclear Engineering at Iowa State University, to convince the Dean of Graduate Studies to sponsor an international meeting on the subject, with financial aid to be provided by the Prince. Later on, additional financial aid was to come from the National Science Foundation (NSF). At that time, as a result of previous work done in industry, I was knowledgeable about the acquisition of images and other scientific information from air- and spacecraft, and was teaching a course in remote sensing techniques. When the dean acquiesced, he asked me to serve on the program planning committee and organize the remote sensing sessions and summary session panelists.

The conference theme centered on the questions, "Is iceberg transport and utilization a viable solution to the ever-threatening worldwide water shortage? What social, political, ecological, and technological problems may be involved in the conversion of a polar iceberg into drinking water? What research must be done? What alternatives are there?" We decided we would seek the aid of NASA, NSF, and foreign scientific organizations to compile a list of invitees and participants. I continue to be amazed by the way experts in things "iceberg" came out of the woodwork in great numbers. Where in the world did they come from, and how come I never heard anything about them? By the time the 5-day conference commenced, we had over 200 attendees representing 20 countries! Can you believe it? We soon discovered that well over 2,000 people worldwide were making their livings dealing with iceberg-associated science and problems! Even the conference "piece de resistance" was impressive!

As the delegates assembled for the opening session at the conference hall, they were greeted by a modest-size iceberg placed in the entryway! It had been flown down from Alaska in the cargo bay of one of the Air Force's largest cargo planes, courtesy of the prince and the NSF! It easily survived the whole week, and was subject to a fancy ice carving for the final conference banquet.

The assignment for the participants under my purview was to determine the feasibility of continuously finding and tracking candidate icebergs of a size suitable for exploitation using remote sensing techniques, and to ensure that they did not contain air inclusions, which might cause them to break up during the long ocean tow. The sought-for icebergs would be of the flat-topped tabular variety, routinely "calved" in the late spring and summer, from the expansive ice shelf that covers most of the Ross Sea in the Antarctic Ocean. The treasured ones would be found floating at the edge of the ice pack. They would be of the 100-million-ton variety: roughly one mile long, a thousand feet wide, and about 950 feet deep. Such an iceberg could supply an adequate fresh water supply to a medium-size ocean-abutting country for a year! It was postulated that icebergs that were much bigger than this would be very difficult to tow, and would probably break up into smaller masses during the better part of a year-long traverse to Arabia.

I had a jump on the iceberg selection, tracking, and air inclusion problem, because I knew from my previous industry experience that, certainly, aircraft-borne, and possibly spacecraft-borne, sensors could detect, image, and inspect for air inclusions in the large icebergs we sought.

I assembled a group of experts to ponder the problems, including a friend who was the head of the Microwave Radiometry Branch at the Naval Weapons Center at China Lake, CA; and experts from the British Antarctic Survey; the Canadian Centre of Cold Ocean Resources Engineering; the U.S. Army Cold Regions Research and Engineering Laboratory; the U.S. Navy Global Ice Analysis and Forecast Office; and a scientist from the Remote Sensing Center of Cairo, who was on a visiting assignment at NASA. The findings of my augmented panel were summarized as follows:

"Remote Sensing methods can readily locate best candidate icebergs, and then, from many, select an optimum at any given time. These methods can also assist in necessary early feasibility experiments by tracking and determining weathering characteristics of 'test' icebergs. Finally, in an operational sense, Remote Sensing methods can assist by observing sea states, recommending best routes, and determining 'weathering' effects during iceberg tow."

The panels that studied the several other avenues of inquiry dealt with problems of towing; erosion of the bergs during tow; effects, once anchored off-shore, on the local ecology, economy and weather; the probabilities of consistently finding suitable icebergs on demand; and the expected costs of retrieval and operation. It turned out that about ten large ocean-going tugs would be required to perform the tow, and that the optimum speed of tow, from the ablation and non-breakup standpoints, would be about one mile per hour. At this rate, despite an estimated 30% loss during travel, there would still be a year's water supply delivered to the Middle East. There was concern that a huge iceberg parked near the shore would

radically affect local weather, but this appeared to be a "you pays your money, and you takes yer chances" tradeoff. As for the economics, there was no doubt that the operation would be expensive but, here again, the potential good appeared to outweigh the cost factor.

My Department was later able to contribute a partial solution to the erosion problem. We had a water channel facility, wherein an overhead carriage could drag an object through a long narrow pool at constant speed. We conducted experiments on identical standard-size blocks of ice. We tried several configurations of porous and solid streamlining bow barriers that were affixed to the forward ends of the blocks. We ran the blocks back and forth several times with the "fixes" attached, and then weighed the specimens, comparing the weights with the bare block results. We found that such barriers could significantly reduce the erosion. A small triumph!

Following the conference, the Prince and some of his colleagues formed the "Iceberg Transportation Company," headquartered stylishly, if incongruously, in Paris. Iceberg towing, from the Ross Sea to Chile, was successfully completed a few years later, but the concept was then abandoned. The experience gained in the Chilean exercise indicated that, at that time, the projected cost of long-range transport and water recovery would be prohibitive.

Perhaps it is time to take another look at this fascinating concept, applying today's technology and economic realities? Maybe now we could better handle a Rhode Island–size iceberg, and turn the whole Middle East into lands of milk and honey!

PIE IN THE SKY

A football-oriented college town in the mid-west is like no other. Ames, Iowa, the darling of the cross-word puzzlers, is the prototype. The people are friendly and far more cosmopolitan than, say, the inhabitants of New York. Why? Because they are so well-traveled, and compensate for their reputed lack of cultural proclivities in heroic ways.

Why the travel? To get away from the freezing winter cold at every possible opportunity. In the 9 years we were there, we took advantage of the opportunity to hear every major symphony orchestra in the world, and to see Broadway plays and traveling art exhibits.

The good citizens even had chances to absorb the niceties of New Orleans jazz in the illustrated concert/lectures called "Serenades For Mouldy Figges," by an expert we all know and love. We lived on a cul-de-sac, Jarrett Circle. All our neighbors were good friends. Professor Al Kraft was one of them.

The letter to the *Sioux City Journal* was right to the point: "Once in a while some college professor comes up with a lulu of an idea. It appears that two Profs at Iowa State University, Kraft and Brodsky, have just won an all-time honor in stupidity."

"They are asking public funding from the space agency (NASA) to support such a simple minded project as attempting to prove the feasibility of storing food in space."

The irate writer, Gene Zortman, of Onawa, Iowa, added: "It seems to me that if we are going to ever balance the budget in this country, one good place to begin would be to get people with brainstorms this outrageous off the public payroll."

The gist of what got him so energized had been picked up by the wire services, and soon became an international joke. Even David Brinkley felt free to express his incredulity on the national evening news. And that wasn't all: The legal office at Iowa State University got a registered letter from Hormel, the giant meatpacking conglomerate. They objected to the name for our proposed project: SPAM (Space Preservation Applied to Meat). They even sent a copy of the "cease and desist" copyright infringement letter to the president of the university.

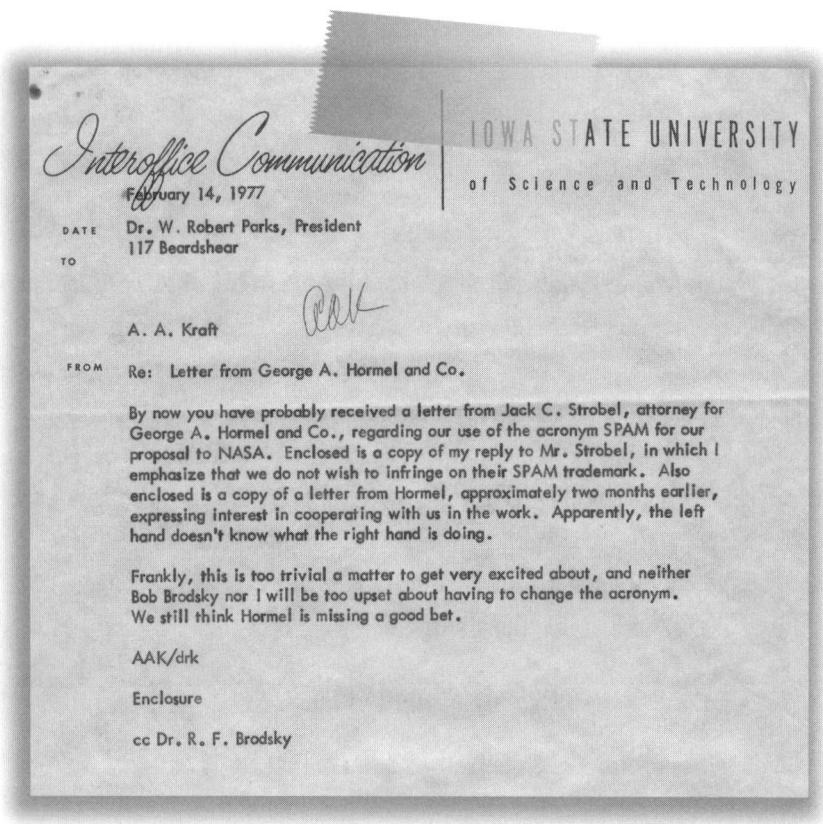

The only known friendly, albeit short term, response that I am aware of came in the December 15, 1976 issue of *The Fayetteville Observer*, from a North Carolinian who apparently saw the wisdom in what we were trying to accomplish. Accompanied by a drawing of a potpie in orbit, their editorial page said, "Credit Allen Kraft and Robert Brodsky with one of the most unusual ideas to come along in the past few days."

The following provides a little background: Our neighbor, Al Kraft, a Prof in the Agricultural College, did his research in methods of food preservation. In the

'70s, preservation by irradiation was the preference under study. Since I was a refugee from the world of intense radiation, as in my prior nuclear bomb design experience, we sometimes discussed his laboratory problems.

In September 1976, I received an "Announcement of Opportunity: Space Experimentation on the Long Duration Exposure Facility (LDEF)," from the NASA Space Technology Office. They planned to have the space shuttle place a very large, cylindrical spacecraft in low Earth orbit for at least a year, after which time it would be brought back to Earth by the shuttle. They would provide room for hundreds of experiments and invited interested parties to make proposals. Enhancing my interest in this Announcement of Opportunity (AO) was the information that the NASA Langley Research Center (near Newport News, Virginia) LDEF Program Manager would be my old friend and prior sponsor, Bill Kinard.

I recalled all my casual conversations with Al, and wondered if he would like to see how meat would hold up after longtime exposure to the cold, vacuum, and irradiation environment of space. He loved the idea, and together we brainstormed a credible experiment that involved several cuts of meat, methods of packaging, methods of exposure, and ways to check the eat-ability, digestibility, and flavor of the returned samples. We proposed feeding the returned samples to dogs first. If they survived, Al said he would be the before-and-after flavor arbiter.

With his department chairman's support, we ginned up an inter-disciplinary proposal, wherein my department would handle the space engineering and integration problems, while his would prepare and evaluate the experimental packages. When we submitted it as a two-year, $130,000 program, we felt we had a winner, despite its quirkiness.

What we didn't anticipate was the ensuing brouhaha. As we quickly explained to NASA, "I am writing you to inform you of some 'publicity' developments with our LDEF Experiment proposal, 'Space Preservation Applied to Meats (SPAM),' in order that you do not misconstrue any publicity that may have received national press coverage recently. A story about our proposal was written by a student reporter covering research projects for the University newspaper. It was made clear to the reporter that this was simply a proposal application to NASA, and that no decision regarding its disposition would be made until early 1977. To make a long story short, the Des Moines newspapers, and then AP and UP picked up on the story and it may have been distributed regionally or nationally."

Shortly thereafter, at our president's suggestion, Al Kraft placated the senior attorney of the George A. Hormel and Co. His letter noted that we appreciated their desire to protect their trademark rights, and that we did not wish to infringe on them. "We will devise a new acronym for the proposed research." Al also referred to an apparently uncoordinated letter from the Hormel director of marketing, which was at considerable variance with their legal beagles. This marketer, in his enthusiasm to inaugurate an upbeat sales campaign, using the idea "SPAM-in-SPACE," suggested testing a sample that included "the red meat ingredients that are used to produce our product," and offered to supply gratis all the samples we could use!

I decided to justify our proposal to the doubting Mr. Zortman by sending a letter to him, with a copy to the *Sioux City Journal*. Painfully lecturing him "so that you won't become a 'know nothing'," I explained that our proposed experiment should provide much needed information for both successful colonization of space and energy savings on long, manned space voyages, as well as for providing means of preserving food for use on Earth, heretofore not considered. I further told him that what we proposed was, in fact, a scientific experiment, e.g., can the bacteria that normally attack meat either be effectively killed or made inoperative in the proper space environment? "As such, from a purely scientific standpoint, this proposed experiment does not really need any further justification other than adding to scientific knowledge."

In answer to his very reasonable suggestion, of using the Antarctic regions for storage, I riposted, "Unlike the South Pole, which is not very cold and has a lot of oxygen, space offers temperatures potentially minus 200 degrees and a complete lack of oxygen in its molecular form. Consequently, it is in theory a much more hospitable environment for long time storage of food stuff."

Now handicapped by a new, eunuch-like acronym, SPOM (Space Preservation of Meat), and already bathed in controversial publicity, it is not surprising that NASA turned down our proposal. In a private conversation, Bill Kinard told me that the sensitivities of government politics, ever alert to ridicule, simply could not allow what he thought was a very good idea to be funded. There is absolutely no doubt in my mind that this experiment will have to be done some time in the near future, especially if a 6-or-more person International Space Station ever gets established.

After reading the *Fayetteville Observer* article, a Carolinian buddy of another neighbor felt moved to send him the following bit of Americana, which he forwarded to us: "Tell them it's all old hat! During WWII, B-17s of the Eighth Air Force were frequently flown to high altitude, 30000 feet or more, for the sole purpose of making enough ice to stock the bar at the Officer's Clubs when big bashes were in the offing."

Everybody got into the act! This beauty from the Des Moines Tribune, *Dec. 8, 1976. And, on bottom, from the* Ames Daily Tribune, *circa same date.*

GREAT COURT CASES #2

One of the first things I noted, after I took my new job as head of the technical staff at the Aerojet Systems Division, in Azusa, was that one guy on my payroll of about 60 people was making almost twice my salary. Moreover, when I inspected the old timer's record, I saw that he only had a Bachelor's degree, in Metallurgy, of all things! So why his high salary? What was going on?

I made it a habit to have head-to-head talks with all the personnel who reported to me. Ted Swanson's turn came early because of my curiosity about the disparity. Our conversation soon unveiled that he had a key, and highly valued skill that few people in the country possessed. He could interpret x-ray pictures of the myriad of welds that were necessary to fabricate all of Aerojet's many high pressure rocket propellant tanks, rocket engines and motors, and other highly stressed structures. He could detect miniscule flaws that an untrained person would never see, and had sole authority to order up rework. He had a remarkable record of success. In the next few years, I had several occasions to consult with him, and he patiently taught me some of the rudiments of interpretation.

Later, in Iowa, this newly acquired capability would come in handy, in a way I could not readily foresee. I got a phone call from a lawyer who described the conditions surrounding the fatal crash of a passenger airplane near the Quad Cities airport.

In the mid-'70s, being the head of the ISU Aerospace Engineering Department—the only such department in the sovereign state of Iowa—I automatically was the ranking local aeronautical expert. It was no surprise to be asked to be a potential provider of expert testimony, in an upcoming case concerning an airplane crash that had taken place in Muscatine, near the Quad cities. I recalled the terrible humiliation I had suffered in a New York City court airplane accident case in the late '40s, but reasoned that I was now more mature, and could better fend off the slings and arrows of a crafty defense attorney. I accepted the assignment for the challenge and the money.

A Muscatine Airlines ancient Twin Beech had taken off from Des Moines with 6 passengers and was heading home when, close to touchdown, its right wing broke off. Everyone was killed; 3 passengers were from Iowa, 3 from Illinois. It was not a totally unanticipated accident. Twin Beeches, which had logged over 10,000 flight hours, had a terrible record of just such occurrences, totaling about 5 identical priors, as I remember. What was surprising, however, was that the accident happened after the Federal Aviation Authority had instigated an edict that required the Twins to be x-rayed every 500 flight-hours, with the purpose of looking for tell-tale cracks in the main weldment that attached the wings to the body. Also surprising was that this plane had been so inspected shortly before the fatal crash. It just had been given a clean bill of health.

The families of the victims quickly opened two separate class-action malfeasance suits, one in Iowa and one in Illinois. The suits were loosely coordinated and directed against both poor, little Muscatine Airlines, which hardly could afford the gas for its two-aircraft fleet, and "big pockets" Beech Aircraft, of Wichita, Kansas, which happily was liability-insured by giant Lloyds of London. The latter outfit maintained a 150- person office in Chicago, which handled liability cases for most of the U.S. airlines and aerospace manufacturers; and, as I soon found out, it was peopled by very sharp "legal eagles," who were expert in all things aeronautical.

To defend their clients, Lloyds of London, joined by Beech's legal staff, had retained a Cedar Rapids lawyer to handle the suit in Iowa. He had a marvelous statewide reputation for winning big ones, and in his wisdom, called me to be his expert structural witness, even though I assured him that my forte was in things related to space. He said he needed the prestige of my position for the trial, and convinced me that, when I looked into the details, I would easily reach the same conclusion of Beech's innocence as he had. The plaintiffs' claim was that Beech had perpetrated a terrible design goof at the junction of the wing and the body, and that they should have known better and done a proper redesign. The rap against the airline was that they were negligent in inspecting for cracks, although it was known that the inspections were done on schedule. The attorney arranged for me to go to Wichita to be briefed by Beech engineers, accompanied by 3 of Lloyds' engineers/lawyers.

Air Force version of a Twin Beech. Courtesy of Google.

In a long day in Wichita, the story behind the crash was unraveled, step-by-step, by Beech engineers. The joint in question was, indeed, a cobweb of complexity, but at the time of the original design many years ago, the fabrication method was state-of-the-art. Six individual structural aluminum tubes were welded together with the main wing spar at the joint. Beech admitted it would no longer use the same design, since it did demand great welding skill. Beech also noted that, in laboratory testing of the joint, after many loading cycles that simulated landing stresses, there was a distinct tendency for cracks to form at the joint. These originally hairline cracks widened and lengthened as the test continued, obviously precursors to the major failures that actually occurred. The test engineers had determined that cracking did not start to occur until after many hours of flight time, which corresponded to many, many landings.

They dutifully reported this phenomenon to the government, resulting in the every 500-flight-time-hour inspection edict, to be conducted once the Twin exceeded 10,000 hours. Beech also prepared an elaborate repair kit, wherein a new structure, overlaying the troubled joint, could be added. However, the cost of buying and installing this fix was about equal to the present worth of the aircraft. We were told that most users had opted for the government-approved x-ray alternative, as had Muscatine Airlines.

When I heard this exposition, I got a sick feeling in the pit of my stomach. How could a responsible company take such a passive approach, especially after the previous crashes? I didn't ask the question, feeling that it was something I should more properly discuss with my employer. I felt that Beech had not been negligent in their original design of the joint, since such joints were commonplace at the time they were put on the drawing board. But I did feel that Beech should have insisted that the fix be applied at its expense, or at least shared the expense with its users.

It got worse! They then showed us the last 4 x-ray sets that had been taken by a firm the airline had hired to perform the required inspection and evaluate the x-rays. They said that the subcontractor had not reported any trouble to Muscatine Air. The minute I looked at the first picture set, taken over 2,000 flight-hours earlier, I detected a small crack in one of the views. I watched in amazement as this crack grew to an easily discernible flaw in the most recent picture set. I concluded that the inspectors either were blind or incompetent, with the latter being the higher priority. I had seen and heard enough and went home to write my report.

I took my long, detailed report to Cedar Rapids. "Settle!" I said, "and settle right now," I cautioned. The lawyer scanned the report and told me that it was not the testimony that he was looking for. "Could you change your report to emphasize Beech's blamelessness?" he asked. I told the lawyer that I had a strong feeling that, despite the fact that Beech had done nothing wrong engineering-wise, I believed that any reasonable jury would find against it with a vengeance. He said to me, "This is a case with big bucks involved. I need a real positive testimony." Mine was obviously not the advice that he wanted to hear, and he thanked me and summarily canned me, with prejudice.

The Illinois case went to court first. The jury awarded the plaintiffs 8 million dollars. My attorney then tried to settle, but it was too late. The Iowa case also went to trial. Lloyds of London had to cough up another 12 million. Muscatine Airlines already has gone out of business. I had been saved from another grueling court experience. Alas, I was, however, deprived of wearing the mantle of "expert witness." The opportunity for such lucrative service never again came up until well into my retirement, when a friend, who had been a patent lawyer at Aerojet, called upon me in cases involving the production of microelectonic circuitry and—Heavens to Betsy!—the design of a golf club, a putter no less.

IN 1984, I "INVENTED" ASTRONAUTICS!

In a now-famous Smithsonian Institute recording session, the fabled New Orleans pianist, Jelly Roll Morton, averred, "In 1908, I invented jazz!" This remark was greeted with some derision by the aficionados of the "hot jazz" genre, because they knew that the music had been already played for several years, and that its most credible founding father is generally believed to be the never-recorded Buddy Bolden. Nevertheless, they did understand what Jelly Roll was really claiming; for it was he who first "wrote" down the tunes on a sheet of music.

In a similar fashion, in 1984, I "invented" astronautics, as a new field of academia seeking both acceptance and accreditation. In achieving this, I—and a few other like-minded space buffs—had to overcome the prejudices of the ruling body of Aeronautical Engineering faculty, who did not want to see fluid dynamics superceded or equalled in importance.

It is my contention that the milepost event that initiated the quest to legitimize "Astronautics" was the pioneering paper "Some Ideas for an Undergraduate Curriculum in Astronautics," which I presented at the 1984 annual meeting of the American Society for Engineering Education (ASEE). In it, I made three points:

1) I predicted that the already-burgeoning field of "Astronautics" would expand to take its place in academia on an equal footing with "Aeronautics"—a field that was, and is, jealously and zealously guarded by the new "Battleship Admirals" of their epoch—the airplane/fluid dynamics devotees;

2) I described 12 "new," specialized courses, added to already prevalent orbital mechanics courses offered by most Aerospace Engineering Departments, which could constitute the basis for either a bachelor's degree or

degree option in Astronautical Engineering. The descriptive listings included heretofore "untaught-for-credit" courses in Spacecraft Design, Rocket Propulsion, Spacecraft Power Systems, Telecommunications from Space, Space Environment, Spacecraft Thermal Control, Spacecraft Dynamics and Control, Spacecraft Instrumentation, etc.

3) I indicated what changes needed to be made to the extant Aerospace Engineering Department accreditation supplementary criteria, to sprinkle the necessary holy water on the idea of a new academic degree program in Astronautics.

The paper noted that the practical problems involved in making this step were twofold: There were no textbooks to cover the subjects, and there were no (or few) fulltime professors with the experience to teach the subject authoritatively. In fact, over 20 years after Sputnik, the so-called "Aerospace Engineering" departments were still woefully unprepared to train students in space technology. The worst part of it was, they didn't know it. By changing their department titles from "Aeronautical Engineering" to "Aerospace Engineering" and offering one or two courses in orbital mechanics, for which there were adequate texts, they believed they were serving their constituents properly.

My paper was the first to make the majority of aero academicians aware of the shortcomings in their curricula. Additionally and indirectly, the impetus the paper provided probably kept the field of Astronautics under the academic rubric of Aerospace Engineering departments, and out of the clutches of other potential university engineering and science departments, such as Electrical Engineering, Mechanical Engineering, and Engineering Physics—all of which might have a legitimate claim to its adoption. At that time, in the early '80s, only the U.S. Air Force Academy—driven by educational needs for its officers that were not being met by civilian U.S. universities—offered courses in "Astro" by a department devoted to the field. Except for the highly mathematical Orbital Mechanics courses, space technology was not part of any other degreed academic program, and thus was unavailable to civilians.

The reason for this is easily explainable; all of the space practitioners were working in industry, or were in the service supporting the military's space programs, or were assigned to teaching at the USAF Academy. In fact, in 1971 when I migrated to Iowa State University, after participating in the exciting initial 13 years of the hectic and mind-expanding space program, I was probably the first early spaceman to enter academia. I had been hired for just that purpose; to spread the gospel to up-coming aerospace engineers.

When I arrived on the fulltime academic scene, I found that because of the lack of texts, space technology courses could not be readily taught by the aerodynamics-trained faculties of the Aerospace Engineering Departments. The knowledge I brought was based on my industry practice, plus attendance at a few early "no academic credit" short courses that had been presented by UCLA Extension in the '60s. I inaugurated courses in Spacecraft System Design, using notes from my industry experience, and in "Principles and Techniques of Remote Sensing," utilizing an obscure book published by the government that did include a few optical and infrared system fundamentals.

Seeing that something more had to be done, I began a long campaign to establish "Astronautics" as a legitimate new academic field. Legitimacy meant changing the rules so that such a new course of study could be accredited—for without bona fide accreditation in a recognized field, industry would not hire a graduate as an engineer. The accreditation procedure for U.S. Aerospace Engineering departments lay in the hands of my technical society, the American Institute of Aeronautics and Astronautics (AIAA). My hope was that with accreditation would come the needed texts.

I attacked the problem on two fronts: curricular and political. In the former arena, I wrote and widely distributed, in March 1985, a white paper entitled "The Establishment of a New Accredited Undergraduate Curriculum in Astronautical Engineering." In this paper I described, in great detail, the core courses that I believed should be included to meet requirements for both the Bachelor's and the Master's degree in Astronautical Engineering. I attempted to get the Space Systems Technical Committee of AIAA, of which I was a member, that had commissioned me to write the paper, to endorse it. This surprisingly did not happen, because the committee was almost evenly divided about the need for a separate degree program; could not agree on a set of "core" courses that would comprise the new degree; and some diehard members would not agree to giving up traditional core courses in aerodynamics and airplane guidance and control. But by sending a draft of the "unblessed" paper to all the aero departments and their Deans of Engineering in the country, I accomplished my goal.

MY WINDMILL TILT

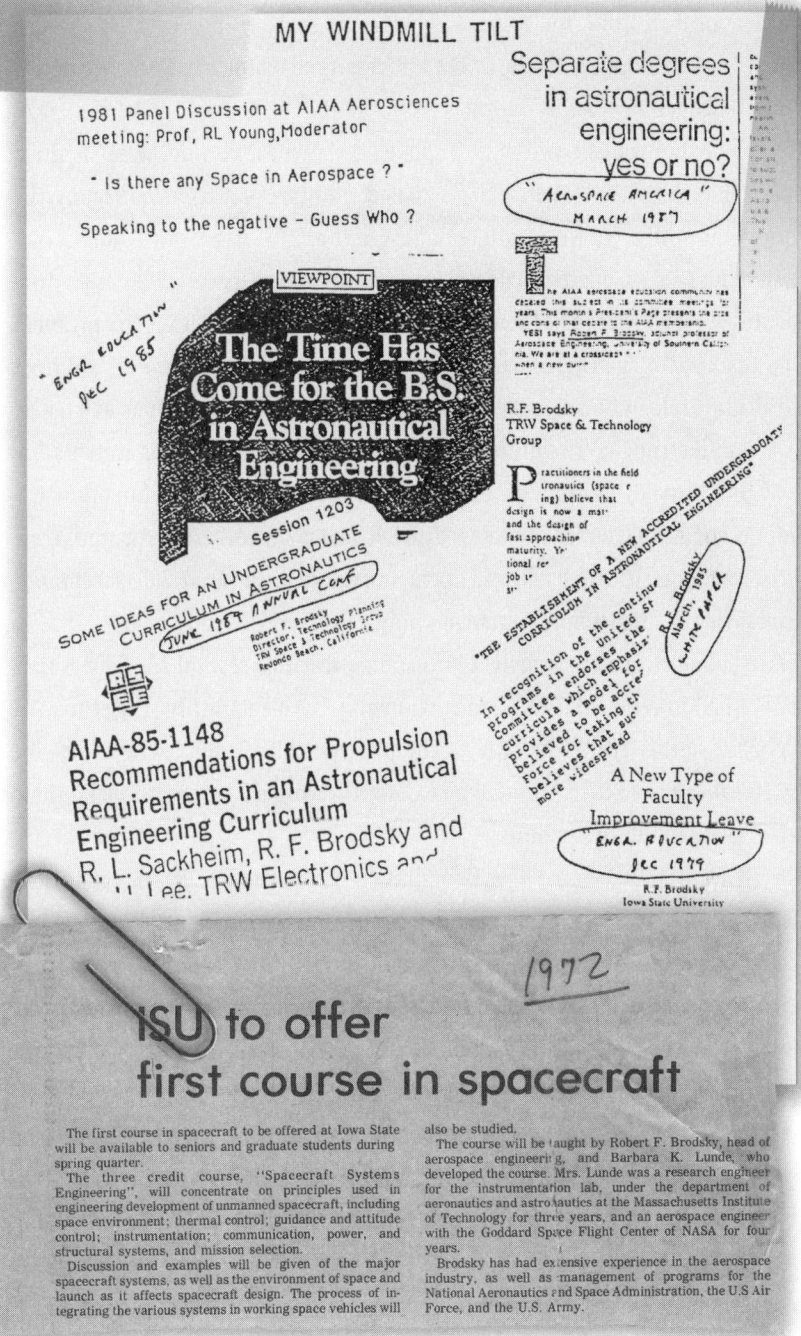

Probably the first "real" space course for academic degree credit to be offered by any school other than the Air Force Academy.

On the political front, the going was just as rocky. Here, however, I was aided by a colleague, Prof. Robert Young of the University of Tennessee Space Institute. In 1981, he organized a discussion session at the annual AIAA Aerosciences Meeting. The panel subject, in which I spoke on the negative side, was, "Is there Any Space in Aerospace?" The debate signaled the start of a reevaluation by the academic community. We were both members of the AIAA Academic Affairs Committee that advised on accreditation criteria, and both of us served as accreditation "visitors," who evaluated Aerospace Engineering departments. The problem was that, except for us, the other members of the committee were all old-time aeronautical people, who saw a new academic field as a threat to their existence. It was an uphill struggle. By cajoling, arm-bending, filibustering, and threatening action by our "Cousin Vinnie," we gradually got the criteria changed to allow the introduction of astro courses into accredited degree programs. This was the straw that broke the opposition's back, and opened the door to the academic acceptance of Astronautics as a legitimate course of study.

During and before the struggle, I wrote and presented several key papers that were published in either the ASEE Journal or the AIAA's monthly magazine. All had the same ax to grind. The title of the paper in 1979 was "A New Type of Faculty Improvement Leave," in which I exhorted aerospace faculty to spend their sabbatical years in the space industry;

The titles in 1985 were "The Time has come for the B.S. in Astronautical Engineering" and "Recommendations for Propulsion Requirements in an Astronautical Engineering Curriculum"; and the title in 1987, in which I took the "Yes" position in an AIAA magazine discussion, was "Separate Degrees in Astronautical Engineering: Yes or No?" My last published gasp came when it was already apparent that the game had been mostly won: In 1995, demonstrating the possibilities of technological progress at the ASEE Annual Meeting, I presented "Teaching Space on the Information Superhighway." In this, I showed that universities lacking the ability to mount an effective astronautics program could have their students take available accredited space technology courses over the Internet, to supplement their on-campus program.

What's the denouement of all this? Well, of the 50 or so aero departments in the country, about 10 have renamed them "Aeronautics and Astronautics," around 15 offer degrees either in Astronautical Engineering, at both the undergraduate and Masters degree level, or degrees indicating an astro (as opposed to aero) option; and most of these offer some space courses, over and above orbital mechanics, which count towards their degree in Aerospace Engineering. And there are now over 200 excellent texts in support of the field.

But believe it or not, some schools still retain the name of Department of Aeronautical Engineering. On the other hand, the program I started at USC is now at the head of the class in the variety of courses it offers in pure Astronautics. This is not surprising, considering that the bulk of the space industry is located within its commuter and web radius, and that USC hires easily available local industry experts as adjunct lecturers and professors. In a personal sense, the best news came in late 2004, when Mike Gruntman, who took over my Spacecraft Systems Design course when I retired as an USC adjunct professor in 1996, sent the following announcement:

> *13 August 2004*
>
> *Astronautics Adjunct Faculty and Lecturers;*
> *Astronautics Program Friends*
>
> *Dear Colleague:*
>
> *In order to position the USC Viterbi School of Engineering to take full advantage of rapidly growing opportunities in space, Dean of Engineering Prof. Max Nikias announced today the creation of a new Astronautics and Space Technology Division (ASTD).*
>
> *ASTD will be an independent academic unit within the USC Viterbi School of Engineering and function in a manner similar to an academic department. The division will be governed by the same rules and policies that apply to academic departments, with its own budget, faculty self-government, and representation at the School's committees and other bodies. I have been appointed chair of ASTD effective August 15, 2004. Aerospace engineering faculty, Professors Daniel A. Erwin and Joseph A. Kunc, join me as the founding faculty of ASTD; several other research faculty and staff will also be a part of the division. ASTD will be supported*

by a new staff member responsible for administrative and student matters of the division.

The new division assumes immediate charge of degree programs in aerospace engineering (astronautics) and of 24 courses to be transferred from the Department of Aerospace and Mechanical Engineering. ASTD will offer a program in astronautics and space technology concentrating on meeting the educational and research needs of interest to the space and defense industries and government research and development centers.

The creation of the new division clearly shows the commitment of the USC to further advancement of educational and research programs in the area of astronautics and space technology. We are excited about this opportunity and look forward working with all of you to achieve the goals of ASTD.

Ad Astra!

Mike Gruntman
Professor of Aerospace Engineering
Chair, Astronautics and Space
Technology Division
University of Southern California

Did I "invent" Astronautics education? I leave that for you to decide. But I defy anyone to find documentation that precedes my 1984 paper, which I believe first sets down a "straw man" curriculum content.

ODE TO TEACHING

When I was young—in my 30s and 40s—I used to act in "little theater" productions, usually playing second banana roles. But I remember clearly the thrill of being a "ham," and having the paying customers hang on to your every word. So it is in teaching. You are the star and you wield major power over your captive audience. It is an ego trip par excellence! And, if as I did, you are teaching to a nationally broadcast TV class, you can pretend you are a movie or TV personality, and that the audience is greatly appreciative of your star performance. You're extra careful to be on your best behavior (no nose picking!), yet urbane and, if possible, witty.

Ever since I retired from the classroom, I continue to get favorable "flashbacks" from my professorial career. All such epiphanies convince me that teaching —no matter at what grade level or environment—ranks among the noblest of all of mankind's endeavors. Particularly if it is done well, or even middling well, the mere act of trying to imbue knowledge and set a civic example are crucial to the lives that are touched by the act. And even to be a "good" teacher, if not a "great" teacher, connotes a pinnacle of success. Alas, in 25 years of teaching at the full professor rank, I only got to the "good" level, but even so, this was a tremendously satisfying plateau of achievement.

The problem of teaching at the university level is that there is no formalized preparation available to assimilate the tricks of the trade. Teachers of K-12 grades have taken specialized training, most of them at colleges dedicated to training students in the art of pedagogy. Not so for teaching engineering or the humanities. Teaching assistants, graduate assistants, and instructors are gleaned from the ranks of those either just graduated at the bachelor degree level or who have a year, at most, of graduate school as a full-time student. Nobody instructs them how to teach effectively; they must just feel their way. And that, my friends, is why there are so many lousy university professors, although hardly any think they are anything less than "very good" or "wonderful" teachers. In fact, I never met an engineering professor—except me—who didn't think he or she was the "cat's pajamas" at the teaching game. On the other hand, as I review in my mind the scores of teachers I was exposed to in my 9 years of undergraduate and graduate education, I can count only 5 or 6 who were truly super—worthy of being called giants!

They had several common characteristics beyond complete mastery of their subject material: They were lucid; they were inspiring; they loved what they were doing, and it showed in every nuance of their delivery. They had "lesson plans" that had been honed from years of experience. They could both frighten you and cajole you to greater heights. They conducted research programs in their field, and thus were able to teach the latest advances. They delivered lectures in a flowing, beautiful, authoritative voice. That's what I always wanted to be able to do when I "grew up"—but, alas, fell somewhat short. I conclude that great teachers are born, not made!

I eased into university-level teaching after my first year of grad school at NYU. I became an instructor in thermodynamics in the Department of Mechanical Engineering. I felt my way into the profession, aided immeasurably by the sympathetic eagerness of the students. They, like me, were almost all returning GIs, whose aim was to get through school as quickly as possible and, 2 to 3 years behind schedule, start a life. Their attitude left a lot of room for learning the teaching game by making mistakes; it was probably the best possible time and circumstances for teaching. Once, when asked a question I could not answer with certainty, I hemmed and hawed, feeling drops of cold sweat rolling down my sides from my armpits, and tried to "BS" my way out of it. The students attacked me unmercifully, recognizing that I was blowing wind! So I learned Lesson 1: If you don't know, tell 'em you'll give them the answer next class. But because of their uncomplicated goal of learning everything thrown at them, I did manage to succeed in teaching undergraduates. Twenty-five years later I was to find out what a fluke that period was, and how really hard it is to teach undergraduates under normal world conditions.

The next University teaching I did was for UCLA Extension, in the late '50s

Pictures from my stint at ISU. Top: Me with model of a gas turbine engine donated by Pratt and Whitney.

Bottom: *Professors Hermann, Brodsky, Iversen and Dean Boylan at the ISU Aerospace analog computer.*

early '60s. Here, I taught graduate level courses, and I did much better. This was because grad students, unlike undergrads, do not have to be led by the nose, and can be depended upon to do their homework and read their texts or handouts. Their careers are dependent on keeping up with the technology. In this environment, since I knew my business, I could take them through the more difficult concepts and could skip the easy ones. Classes were more like clinics on special topics. I done good!

I found out about the inadequacy of my ability to teach undergraduate courses when I joined fulltime academia at Iowa State University. My previous stint at the graduate level left me with a style that reflected my UCLA success. I soon found out that my techniques just didn't work at the lower level. Everything had to be explained in petty and logical detail. I could not depend on the students to read the text or do the homework. I could not depend on them to be able to interpret my notoriously illegible "hen scratchings" on the blackboard. I somehow managed to put the bright ones to sleep and befuddled the average. It took me 3 years to become "adequate," and I never got much beyond that level of achievement. On the other hand, to my amazement and chagrin, I sat in to observe classes being conducted by the reputed good teachers on my faculty, and quickly noticed the difference. My God, one of them could even talk on a related subject while writing pertinent complicated formulae on the blackboard! Surely, one must be born with that facility! It just wasn't my dish of tea.

Because of my administrative duties, and not conducting research programs or consulting with industry, I did not have the smarts to teach at the graduate level. So I struggled through 9 years of befuddling undergrads in the new field of astronautics. This all changed during the 15 years that I taught as an adjunct professor at USC, in the '80s and '90s. Here, I was either working fulltime at TRW or, after retiring in 1988, consulting in industry, until I fully retired in 1996. I knew what I was talking about. I was living it! I taught two different graduate level courses, one on spacecraft design and the other on space-based remote sensing systems. I was at the top of my game for most of this period. I taught over a closed circuit TV network, although there were "live" students in the studio. My notes had been pre-prepared and distributed—obviating the need for blackboard work. In my evenings at USC, two classes a year, I taught a lot of students: my classes averaged 50-60 attendees, so I affected at least a thousand students. As time went by, I morphed into a "very good" teacher. Alas, "greatness" continued to elude me, but I felt satisfied with and pleased by my efforts.

And these efforts were not in vain. During my career, I received two national outstanding aerospace teacher awards. However, I hasten to note, these awards were not for being an exemplary classroom instructor; rather they recognized my "public relations" efforts in trying to establish "Astronautics" as a legitimate university curriculum and a new career opportunity. In 1978, I received an "Educational Achievement Award," jointly given by the Aerospace Division of the American Society for Engineering Education and the American Institute of Aeronautics and Astronautics. This award was for my work in successfully promoting aerospace education at a time when enrollments were at an all-time low. In 1979, I was named the "University Professor of the Year" by the American Society for Aerospace Education," for encouraging and promoting the teaching of aerospace–related subjects in both junior and senior high schools. But, as can be imagined, these awards for writing papers and lecturing to increase public and engineering awareness had nothing to do with improving my classroom performance.

In 1996, I perceived that I was becoming obsolete technically, in the fast-moving remote sensing field. (What the hell, the electronics that were so secret and fabulously expensive 10 years prior could now be purchased in a camcorder for $400.) That fact more than anything else, even though I was still in the best of health, led me to quit the teaching profession while I was ahead, and I did.

But I walked away from academia firmly believing in my opening premise: Teaching is a noble profession and ranks with research medicine (and politics, as practiced by a miniscule few) as one of the finest things mankind can do to improve the general condition of the world's population.

Chapter 6
BACK IN OUR OWN BACKYARD—TRW
1980-88

RETURN TO INDUSTRY!

In the academic year 1978-79, I spent 10 months at the Hughes Space and Communications Division in El Segundo, close to the LA Airport. Having been away from industry for 7 years, I approached this experience with some trepidation. The fear of technical obsolescence in the aerospace business is a real one. I found myself surrounded by bright, young people who were routinely applying advanced techniques to ongoing projects.

The sabbatical experience, as the first of a series of exchange professors, had been exhilarating. I found that I could, indeed, compete technically with the young bucks, and liked being a "field hand," as opposed to an administrator. I also realized how much I missed the excitement and competition of industry. The family, too, recalled how much we missed California and year-round sailing, and how nice the South Bay area was. Upon my return to ISU, I discreetly inquired about a regular job at Hughes, but they said that since they had now created an annual "Visiting Professor" program—so pleased were they by their experience with me in my just-completed year —that hiring me would make it look like a recruitment ploy. I scratched around and found a good "Senior Systems Engineer" job at their archrivals, the equally-noted and close-by TRW. Both organizations were at the top of the list of those who could produce the most sophisticated civilian and military satellites.

I started working at TRW in August 1980. I knew I had the option to go back to Hughes after a decent interval, but it turned out that I enjoyed the TRW action, and stayed there until I retired from a fulltime job in industry. I also had the opportunity to resume teaching my space courses at the graduate school level at night, at the University of Southern California. I started doing that in 1982 and continued until I retired from USC in 1996.

We moved to a town house in Hermosa Beach, two blocks from the beach, 4 blocks from King Harbor, and 3 miles from work. Soon, I was co-partner in a Catalina 27 sailboat, and began sailing twice a week, a ritual I maintain to this day.

THE SHUTTLE BUS CAPER

I attacked my new assignments at TRW with great enthusiasm. At the outset, I was put in charge of a research project that turned into a NASA study contract, "The Shuttle Bus." I next led a winning proposal to the Jet Propulsion Lab on a "Mars Explorer" spacecraft and directed the subsequent study.

Then, spotted by a Vice President who would become my next boss, I took on a prestigious and better paying high level group staff job: Director of Technology Planning. I rued leaving the trenches, but the inducements—like an upper floor, corner office with a couch in it—won me over. This change put me in charge of approving and overseeing all company-sponsored research, as a "side" job. My serious charge was to, somehow, make sure that TRW would never be taken by surprise by a competitor coming up with a pertinent, advanced technology that we were not aware of and could not respond to.

I also started teaching the two astro undergrad courses I had developed at Iowa State, but now at the graduate student level—"Spacecraft Systems Design" and "Principles And Techniques Of Remote Sensing"—at the University Of Southern California, where I was appointed a full Professor.

While on staff, I also got an assignment that nobody else wanted. It was to head up a "rainbow-like" committee to oversee the group's reponse to the new and rather rigid government-imposed rules concerning a new dictum —"Affirmative Action." It had some funny consequences that added to the spice of life.

In the early '80s, the Air Force operated a few key, highly secret satellites that had to be placed into the 22,000 mile-high Geo-synchronous stationary orbit, in the plane of the Earth's equator. At this altitude, their orbital speed is exactly the same as the Earth's rotational speed, making them appear stationary in space. At TRW, we knew that the Air Force would soon have to replenish its dwindling supply of the earlier stockpiled rocket stages, which powered these satellites from their low altitude Earth-circling orbit to the Geo orbit; and that there would be a lot of money involved for the winner of the forthcoming competition. I was put in charge of an R&D program aimed at providing TRW with a winning candidate. I failed magnificently!

My first assignment at TRW, in August 1980, was to take over a company-sponsored independent research program aimed at the development of a unique low cost, upper-stage rocket booster, called the Shuttle Bus. It was designed to deliver a heavy satellite from a low, 200-mile altitude, say, Earth's circular orbit, to a geostationary circular orbit in the Earth's equatorial plane. When I took over the project, much of the preliminary design work had been done. A crucial, yet trivial, rocket engine—"proof-of-the-pudding"—test had been tentatively scheduled. The company then had to make a major decision: should we build a demonstration model with our own money, or find a government sponsor to foot the bill? In retrospect, although I had lots of help in doing so, I think I really screwed up on what, as time went by, I began to believe was an outstanding solution to a problem that still besets us. Here's the tragic story:

While I had been on sabbatical at Hughes two years earlier, I thought that its engineers had found the best solution to the problem of getting a satellite from low Earth orbit to geostationary orbit. It was first applied on its LEASAT satellite which I had worked on during my faculty leave. It carried a massive, very expensive, solid rocket motor integrally inside the satellite. This high thrust motor propels the satellite from its low altitude orbit into a large, elliptical orbit, whose peak altitude, or apogee, was close to the needed 22,000 miles. After it had accomplished its propulsive task, the empty case of the burned-out motor was ejected to rid its weight from its host.

Then, a series of much smaller liquid rocket engines, called thrusters, were sequentially fired at both apogee and at the low point of the elliptical trajectory, called perigee, to finally change the elliptical orbit into the desired 22,000 mile circular orbit. As many as 3 or 4 complete elliptical orbits, each taking a few hours, were required before final circular orbit was achieved. The elliptical trajectory picture thus looked like the path of a yo-yo whose string length was increased each time it whizzed by the holder's hand.

Artist's sketch of shuttle bus with satellite payload on the way to geo orbit. Note 4 large propellant tanks and 4 small thrusters (mounted on top fuel manifold), each with about 100 pounds thrust. Courtesy of TRW.

Although the cost of the large solid rocket motor was very high, the low thrust thrusters were inexpensive, and certainly much cheaper than using a second solid rocket motor to do the job. We knew this because the Air Force, via Boeing, had developed an upper stage to do this job, using two large, solid rocket motors, each costing millions of dollars. This booster stage started out by being called the

"Interim Upper Stage," to connote that they were sure something better and less expensive would soon be coming along. When that didn't happen, they decided to call it by its present name: the Inertial Upper Stage, or IUS. Everyone knew that they would gradually use up their initial buy of IUSs. They had purchased them at about 40 million dollars each. Our strategy, if all went well, would be to propose our low cost, but untried, Shuttle Bus in the competition, hoping our low price would overcome the Boeing offer of an up-dated version of its tried-and-true IUS stage.

When I first took over the Shuttle Bus R&D program, I was still enamored of the Hughes approach, and secretly looked down on our candidate as an inferior design. I'm afraid this may have warped my judgment. The reason the TRW shuttle bus stage would turn out to be inexpensive was that it only employed small, 100-pound-thrust liquid rocket engines, compared to the many thousands of pounds of thrust delivered by the very expensive IUS and LEASAT-like large, solid rocket motors. These small engines would be variants of the thrusters used on the Space Shuttle, and thus would cost only a "dollar a dozen"—relatively millions versus hundreds of dollars!

The Shuttle Bus was a separate stage, to which the satellite to be delivered to high orbit would be attached and then separated from, once on orbit. In this respect, it was just like the IUS. It consisted of 4 equal-size tanks, to carry the high volume load of the liquid propellants: two fuel tanks and two oxidizer tanks. It would be powered by either two or 4 of the small thrusters adapted from the Space Shuttle.

We were about to make up our minds on the number by making a tradeoff between the cheaper two-thruster version, which would take longer to get to final orbit, and the 4-thruster version, which would be more expensive and complicated, but would achieve orbit many hours before the two-engine version. The bus and its passenger satellite would start in low Earth circular orbit and burn its engines for about 10 minutes. This would turn the circular orbit into an elliptical one, with the low point nearest to Earth, being where the engine burn took place. At each subsequent passage past this low point, 10-minute burns would take place, sending the high point of the trajectory higher and higher. After 15-20 of these burns, the high point, the apogee, of the trajectory would finally reach 22,000 miles.

At this point, the thrusters would be fired for about 20 minutes, 10 minutes before reaching apogee and 10 minutes after passing the high point. Three or 4 of these burns would finally round out the trajectory into the desired circular orbit at

22,000 miles. Because the efficiency of the liquid rocket engines was considerably better than that of the solid rocket motors, and the densities of both liquid and solid propellants were about the same, the Shuttle Bus should turn out to be considerably lighter than the IUS, as well as cheaper. On paper, this looked very favorable to us. The only objections the Air Force could have had were the obvious lack of flight test proof and the equally serious additional time it would take the Shuttle Bus to place its satellite in GEO, compared to the approximately 8-hour IUS time to final orbit.

We could do nothing about the latter, but I felt we could do something about the former. I had limited funds available to me. I had set some aside, to have the makers of the Space Shuttle thrusters test their engines, using a slightly different fuel formulation than was being used in the Space Shuttle application. I thought the test would be a "slam-dunk," and cancelled it in favor of using my remaining funds to initiate a detailed cost exercise, wherein I paid various company specialists to determine how much it would cost us to design and build a prototype, which the Air Force said they would test at no cost to us.

It turned out that a reasonable figure to do both was about 15 million dollars. This figure really knocked me out. I finally began to realize the truth that my predecessors had already known. The Shuttle Bus idea was a powerhouse! I was turned into a zealot, and began a quiet (aye, there's the rub) campaign, with the powers-to-be, to get the company to kick up the 15 million and build the damn thing. Well, I simply didn't yell loud or long enough, and probably not convincingly enough. The company hemmed and hawed and contemplated its navel. I let it happen. I know now that I should have "burned down the barn" to get the proper attention. By not doing so, I deprived my company of both a long-term "cash cow" position and a solid place in the space firmament.

Several months later, when the Air Force put out a Request For Proposal for an initial buy of 40 units, with another 40 a possibility if the price were right, we did propose the Shuttle Bus at a fixed price of around 11 million a copy. This would still have given us a fat profit, if we ran into no pitfalls.

Inevitably, Boeing again won the follow-on order, this time at 20 million each. When we asked the Air Force why it turned down the low bidder, it debriefed us: "You have done nothing sincere to show you believe in your approach. We simply couldn't take a chance on a concept with no development behind it. Hell, we didn't even know if your thrusters would work!"

BEATING THE FOG FACTOR

I spent a great deal of my adult life in industry, writing and managing proposals for new work—work that usually was at the cutting edge of technology. My proposal writing style was different from the normal staid, dry approaches that people thought were *de riguer*. It was more intimate and breezier than the usual engineering prose. Some people took offense at it, even though it consistently resulted in wins.

Gene Spangler was the *major domo* of the proposal organization that supported the chase for new work at TRW. He supplied the office space, editors, equipment, artists—all the accoutrements needed to turn out winners. He was management's first line of defense at quality control. We developed a good relationship over the years. Here's how it started in the early '80s

---

In line with Congress' demand to reduce the cost of planetary exploration, NASA—in a precursor mood akin to the '90s' "Faster, Better, Cheaper" environment—wondered if already operational Earth orbiting satellites could not be adapted to perform lunar and planetary missions. NASA's Jet Propulsion Lab wanted to make a test case of this idea, on a new program that would ask satellites to circle both our Moon and the planet Mars, from north to south, going over its poles. We had a production Navy communications satellite, the Fleet Satellite Communications, nicknamed FLTSATCOM, which orbited the Earth in the equatorial plane. I thought it could be readily adapted to do the Mars and lunar jobs.

My proposal opened by postulating, "The selection of the FLTSATCOM spacecraft for the Mars Geoscience Orbiter (MGO) and the Lunar Geoscience Orbiter (LGO) missions permits TRW to offer a proposal which meets JPL's twin goals of

finding a common existing economical bus to perform both missions with comfortable margins and a bus which will remain in production during the critical years of MGO/LGO development. The FLTSATCOM is near perfect because:
• It is a communication satellite.
• of its long duration life on-orbit, etc."

In retrospect, one can see that the opening sentence is a doozie! The proposal prompted Gene to declare in Proposal Management, Issue #60 that, "This has a fog index somewhat over 14. It can be easily revised to 11." The latter number was a minimum goal to strive for.

He then printed his suggestions for an "improved" beginning. Next, one of Gene's henchmen, in his class on proposal writing, further pontificated that, "This is not a winning proposal."

An Honest Tale Speeds Best Being Plainly Told was the title of Issue #60, in the May 1982 edition of *Proposal Management*. It discussed the "Fog Index." It was edited by my friend Gene Spangler.

> "*In fact some years ago, in trying to deal with bureaucratic density, a hopeful communicator invented the fog index to help writers decide whether their writing was good or bad.*" Deciding that most bad writing resulted from big words and long sentences, that inspired person made up a formula:
>
> Fog Index = 0.4 (average length of sentence + % of big words)
>
> A big word is defined as any of 3 or more syllables, or a short-word compound like "everything" or "bookkeeping." The scoring is ranked as follows:
>
>   2 - 8 simple
>   8 - 14 acceptable
>   10 - 12 ideal
>   14 - 17 too ponderous
>   over 17 unreadable
>
> In general, the lower the score the better. The nation's largest daily paper, the Wall Street Journal, got that way by lowering its fog index to 11. Time and Newsweek also average 11."

Being over 14, this was too much for me to bear! I promptly drafted the following letter, reproduced faithfully, although, in the interest of keeping down the fog index, not in its entirety:

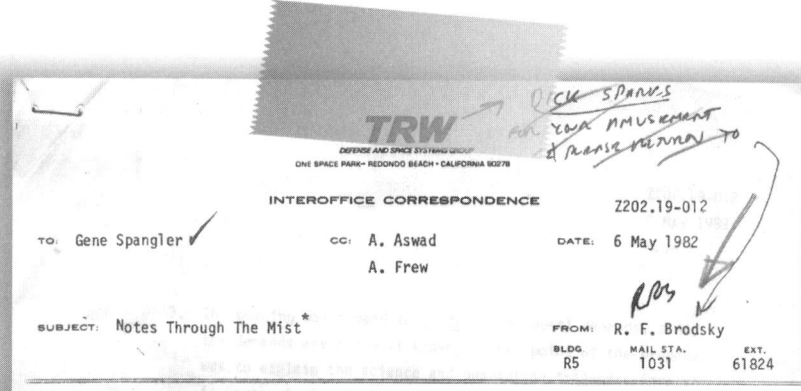

TRW
DEFENSE AND SPACE SYSTEMS GROUP
ONE SPACE PARK • REDONDO BEACH • CALIFORNIA 90278

[handwritten: Rick Sparks — for your amusement & transmittal to]

INTEROFFICE CORRESPONDENCE   Z202.19-012

TO: Gene Spangler ✓   CC: A. Aswad   DATE: 6 May 1982
                                 A. Frew

SUBJECT: Notes Through The Mist*   FROM: R. F. Brodsky
                                        BLDG.   MAIL STA.   EXT.
                                         R5   1031   61824

Were it not for my innate sense of security and Joie de Vivre, I might feel that "they" were picking on the new kid on the block. First, Mr. Aswad denounced (in his class) my proposal as "non-winning," (1) and now Thou, the Father of all proposals, has viciously turned on me. All right--I admit to being a "14"--It's part of my mystique. However, I would like to point out why, in turning a "14" into an "11", some gentle, beautiful, and (possibly) important, thoughts might get lost.

I have done this by marking up old "#60" and providing annotations, as attached. You might inquire as to why I write such notes as this:

1. It is **not** for vindication, since my ego is enormous, and I am too ignorant to understand when someone is trying to help me.
2. It **is** because I like to write whimsical notes--the high points of an otherwise prosaic engineering career. So, enjoy, and take heed, but do **not** take offense.

ANNOTATIONS

1. By saying a "communications" satellite, I hoped to get across several notions: The payload was a communications one--implying heavy weights, large power demands, and considerable heat dissapation, as well as having good stability and pointing capabilities In short, that adjective says a lot to a spacecraft buff.
2. This comment -- "Difference between a long life and long duration life" is what we, in the literary set, call specious.

RFB:BEM

[handwritten: Ego intact along with sense of humor. I'm glad of both   Gene]

Gene sent my memo back to me with the note, "Ego intact along with sense of humor. I'm glad of both." He also referred me to the last paragraph of #60: "Writing is meant to be communications. That is best done simply and briefly. It's easy. Just keep the average sentence under 20 words. Never use a big word unless you must (e.g. never write "presently" when you mean "now"), and, most important of all, write exactly what you mean, not an approximation that forces the reader into guesswork and independent decoding." I am sure Hemingway knew this. Alas, I never took it completely to heart. I like to use big words. They are nice and make the reader look them up if they're confused. I like the reader to use imagination and thought in trying to decipher the meaning. But, it's true—Papa did make the big bucks.

There is a sordid end to this story. We completed the Phase 1 study, devoted strictly to Mars exploration. For reasons of TRW management direction, we dropped out and did not compete for Phase 2. Eventually, NASA selected a contractor, who basically had to build a new satellite from scratch, thus loosing the reliability that comes with a tried-and-true design. Perhaps TRW management got the word of this change of heart "under the table?" Or perhaps the FLTSATCOM, at around 40 million a copy, was thought to be too expensive to start with. The program was renamed "Mars Observer." In mid-1999, following its launch, the "new" Mars Observer disappeared, either in or on the vicinity of Mars, and was never heard from again.

GOOD COP, BAD COP

As Director of Technology Planning, my assignment was to make sure that the research work being done throughout the company was such that we would always be "on top" of the highest technology out there, and that we would never be bested by some new, unforeseen technology breakthrough. In other words, try to find a crystal ball that would predict the future. Then make sure we, TRW, would be ready for it! At the onset, I thought this assignment would be a difficult one to fulfill, but not impossible. It took a year or two to reach the conclusion that I was probably wrong.

Another bonus that came with the new job was an extra room to house my fabulous lend-out technical book and paper collection, which I dubbed "The RF Brodsky Memorial Library and Reading Room."

As part of my new job, I became responsible for the Group's IRAD (Independent Research and Development) program. Each year, the government provides funds that permit aerospace companies to undertake approved research projects. The amount of money provided is a percentage of the past year's gross sales to all government agencies, and also accounts for the government-graded performance results of the prior year's research program. In TRW's case, this was a wad of money—in the multi-millions—vied for by the various technical groups that wanted to investigate new approaches. So, in addition to my function of assuring that we were never caught with our technical pants down, I was also charged with the administration of the IRAD program. Either one of these functions could have been a fulltime job, depending on how I chose to prioritize them.

Moreover, my boss soon asked me to take on another responsibility. The new philosophy of "Affirmative Action" was coming into vogue, and was being taken very seriously by the company. Minorities must be found, hired, and nurtured! The government had an effective "carrot and stick" grip, by using compliance as a factor in contract letting and IRAD funding. I was asked to organize a committee to oversee the progress in the Administrative Staff Group. My greatest subsequent contribution in this area (if you don't count the sailboat ventures) was to fight for a TRW-operated day care center on or near our campus, so that women, minority or not, could have a better chance at holding a good job. The top financial VP vehemently opposed this move, and it was not until after his death, and my retirement, several years later, that the center opened. It is now one of the best in the business.

But I was busy—very busy—trying to figure out how to do my primary job, and then, after I found a possible way, doing it. I asked my boss if I couldn't hire people to assist in the IRAD administration. When he acquiesced, I opened a Pandora's Box that was at once great fun and a great ordeal, and demanded summoning up all my past-honed people skills.

As I took her on to run the formalities of the IRAD program, Nan Glennon was a handsome woman, given to lacy frocks and big hats. It was no stretch to see that she must have been a gorgeous colleen in her younger days. She was the first woman engineering graduate of USC, and had had many variegated jobs at TRW, over a long period, including a spell in the Washington office. Her qualifications were undeniable. She had fought her way up the ladder, helped no doubt by her being tough as nails, and certainly not hindered by her good looks. We quickly developed a love/hate relationship.

She loved me for the power that I delegated to her and for my generally backing her in the many disputes that arose with her clients. She hated me because I was always telling her to ease up and "letting those smart-ass engineers get away with murder," when I had to mediate before blood was spilled.

I loved her because of her ability to establish a benign reign of terror—that made normally slothful engineers turn in both their programs and progress reports on time—and her amazing attention to details. She read every line submitted with care and thought. If a comma were missing, or if a sentence didn't really say what it meant, she would bounce the report and demand an immediate corrected return. On the other hand, this attention to details was one of my weakest areas; I always went for the bottom line and cared little for the small print.

I hated her for the time I had to spend smoothing the riled waters, both with her clients and then with her. Almost daily, in report season, I would get an angry call, "Get that blonde bitch off my ass!" Sometimes, when they thought that her rewrite demands were just too unreasonable, they would storm directly to my boss, and he would have to call me in and hold my hand. So, I often had to rein Nan in. She bore no grudge, and forged on to the next case. In all, I really loved her!

Our "good cop, bad cop" modus operandi went on for 3 years, and we consistently won both "best in class" awards and lots of government award money. Nan was solely responsible for this record. I was only concerned with the type and quality of the research. My boss was happy because I was a buffer that kept him out of the fray, and we produced excellent results.

But too much of a good thing was wearing on me. I yearned to get back into the new proposal business and to writing, and then presenting, technical papers in far away places. Probably consciously, I gave a paper to the 1984 meeting of the International Astronautical Federation in Lausanne entitled, "Technology Planning in a High Technology World—Can it be done Successfully?" The paper summarized my attempts to lead the TRW research efforts into high impacting, heretofore unforeseen, new areas. I reported convincingly in the negative. "The job simply couldn't be done effectively," so I thought and so I said.

Upon my return from Switzerland, I stuck a copy of the paper under the new big boss' nose. A few days later, I was transferred back into the line as a ranch hand, and my directorship was abolished. My former boss inherited Nan, and retired shortly thereafter. I missed my duels with Nan and the Affirmative Action Committee work, which was always a source of fun in an "Alice in Wonderland" way. But I was, again, gainfully employed until my retirement. As a postscript to this tale, in December 2005, I was sorry to read of Nan's passing, in the TRW Retirees Bulletin.

TECHNOLOGY PLANNING IN A HIGH-TECHNOLOGY WORLD—CAN IT BE DONE SUCCESSFULLY?

R. F. BRODSKY

TRW Space and Technology Group, Redondo Beach, California, U.S.A.

Abstract—An important problem facing high tech enterprises is finding the means to ensure that they will have on hand the technology needed in the future to maintain their present competitive advantages. The paper describes the way three organizations—the U.S. Air Force Systems Command, the NASA headquarters research organization (OAST), and a private company—have addressed the problem. Some technology extrapolations are shown; however, the author has reservations about how such future requirements can be translated into high-leverage organization technology acquisition. A formalized process does yield long-range goals and heightens awareness of future requirements. The difficulties arise because there are many more legitimate goals than available funds to acquire them.

INTRODUCTION

In the past few years, it has become obvious that the burgeoning advancements in microelectronics, computers, electro-optics, and microwave technology have or will open avenues heretofore unavailable or even unthinkable. In the limited economy environment that governs any endeavor—be it a government, a governmental agency, or a private business—the problems quickly reduce to finding some near-optimum plan of research and development funding. The next level of planning requires the division of such funds into those needed to support near-term objectives, while most judiciously applying and protecting those funds required for future growth and competitiveness. It is the latter problem that provides the subject matter of this paper. The funds available for long-range R&D turn out to be precious few. In industry, the Independent Research

(1) His own company, the TRW Space & Technology Group, as a result of the establishment of a Technology Planning office charged with the responsibility to ensure the Group's future competitive ability as well as to operate the Independent Research and Development (IR&D) program, and to facilitate technology transfer, chiefly between Space and Technology and the other three high technology Groups (Electronic Systems, Defense Systems, Energy Systems) which comprise the noncommercial portion of TRW's Electronics and Defense Sector.

(2) The Systems Command (AFSC) of the U.S. Air Force, via the AFSD (Air Force Space Division) and the AFSTC (Space Technology Center), as present chairman of the Future Concepts Panel of the continuing MSSTM (Military Space Systems Technology Model) Workshop III effort, and past participant on the Future Concepts and Operations Panel of Work-

---

Interoffice Correspondence
TRW Space & Technology Group

P700.3-022

Subject: Some of these days, you're gonna miss me, honey!

Date: 19 January 1983

From: R. F. Brodsky

To: Distribution

cc:

Location/Phone: R5/1031 61824

Friends, Romans, Countrymen - before ere long, I and my beloved lending library will be leaving my present vale of tears, Room 1031 - perhaps to gaze out of a window in some finer* place on the campus where I surely belong. To all of you, my equally beloved clients, I ask - indeed plead - return my borrowed treasures! I think, longingly, of my Hughes IRAD report comparing spinners with 3-axis spacecraft (Sabroff, Frew, Love?), of "Remote Sensing of Earth from Space" (Nardone?), of the AIAA "2020" green book (Hieatt), of "Scientific Satellites", etc. (Sparks), the ESA red-orange volumes on European Space Structures (Waltz?), and many (?) other best sellers that senility precludes me from mentioning. Please, send my "chillun" home!

RFB:eem

* I have asked for an octagonal room to accommodate more bookcases.

OLD CYBERSPACE U.

When I first wrote this story in 1999, the University of Southern California had already achieved what was arguably the broadest selection of academic courses in astronautics in the world. As you will find out, the day of the virtual university that could spread this gospel was in sight. In fact, by 2004 when I made a slight rewrite, it had arrived.

Shortly after I started working at TRW in 1980, I got a call from an old friend, Dr. Janos Laufer, who was professor and founding head of the USC Aerospace Engineering Department. He asked me if I would consider teaching night graduate school courses in my field of space science and engineering. He pointed out that except for the almost mandatory course in orbital mechanics, USC had no other offerings in the now burgeoning space arena.

In the early '70s, at Iowa State, I had originated two new courses, one dealing with spacecraft design and the other with remote sensing instruments. I told him I would be pleased to teach updated versions of these courses at the graduate level, so long as my bosses at TRW had no objection. They didn't, so I did.

For the last 8 of the 15 years that I was an adjunct professor at USC, my classes were presented to the local aerospace companies over an interactive TV network. I enjoyed teaching via TV immensely. I used to be active in little theater, doing second banana and dialect roles. The "ham" in me reveled in having a captive audience to perform before. There is a lot of "show business" in the distant learning game, as well as a coming opportunity to be your own stage manager.

As I was leaving the academic scene, for retirement into an author's life, the USC people who operated the Interactive Educational TV network told me that I was in the vanguard of what is becoming a revolution in the way instruction is being delivered to an on-line public. In the future, they opined that an instructor would sit in his office at home, just as I am doing now as I write this, and deliver lectures for all the world to see and hear. High-speed telephone lines will carry the lectures, in living color and stereo sound. The new look in educational delivery will be similar to the USC studio set-up that I was now used to, except for the absence of live students sitting in front of me. Here's how it will work:

In the home office, a single, 2-position digital camcorder either will be able to look directly at the instructor or look down at a page, such as a pre-prepared page from a book of notes or a page on which the instructor can write. If two camcorders are employed, then superimposition of the instructor's image in a corner of the screen, simultaneous with looking at a page, will be possible, all on command from the instructor. The students will receive the information, broadcast via a microphone, over the telephone system that sits either in their home offices or in an office their company has provided. If a student has a question, simply picking up the nearby phone will connect him or her directly to the instructor, and the question will be heard by everyone in the class—as, of course, will be the instructor's answer.

The student will have a complete set of the pre-prepared notes on hand to mark up, as needed, during the lecture. Examinations will be done either by e-mail in real time or by pre-arranged proctoring services, set up at a local university. Homework, taken from the pre-prepared notes, will be sent in by e-mail, but also will be discussed during "class" time. Corrected homework will be returned via e-mail. As now, the professor will have one or more graduate assistants to do most of the scut work. The beauty of it is that, a registered student, anywhere, can have a selection of courses to choose from, taught by outstanding faculty members at a variety of participating colleges.

That's the predicted look at the future: an electronic university! A place of learning much like today's schools, but with he added feature of real time or delayed distant learning. Will it work? Based on my USC experience and the progress made since I retired from teaching in the late '90s, I think it will. Moreover, I think it will yield a satisfactory educational experience. What will be missed is the intimacy of a classroom, where one meets and interacts with fellow students and directly with the instructor. Still, I did feel a bond with the disembodied voices that interrupted my lectures, either to ask questions or make sometime cogent comments. And I believe we began to know each other through my written remarks on their graded homework, and the brief times we spent together in person at the midterm and final exams.

To illustrate the efficacy of the distant learning process, I can cite a student in Rochester, New York, who turned out to be "top dog" in my Spacecraft Systems Design course—and he learned by recording my lectures and playing them back at his leisure! This convinced me that long distance learning can be as good as the old fashioned way, at least for the dedicated. The point is, that with e-mail, a student can get more of a teacher's time than in the old-fashioned, end-of-class conversation mode.

Modern communication abilities will clearly bring about a gradual change in the way education is delivered, and in the very basis of a university. Students will be able to "tune in" on courses given at many universities. If a lecturer at Harvard, say, is a known expert, and his specialty course is highly lauded, why not have it available for all students—and non students too, for that matter—to take it for credit? Billing can be done electronically. Think what wonderful curricula can be ginned up in the pursuit of a degree!

Such a scheme is presently under way at the graduate school level, sponsored by the National Technological University. NTU grants an accredited master degree in various engineering subjects, by "buying" a large number of courses from many universities that have accredited programs. It is strictly a paper university, but NTU's students are receiving lectures from top people throughout the country—and this approach easily can be extrapolated throughout the world. NTU collects the tuition and pays both the transmitting school and the individual lecturers.

It will take a lot of hard work to make this happen, but the old-line universities need not fear being supplanted. They will still be needed. But soon the bed-ridden and the boondocks can be served. And, most impressive, imagine being able to be inspired by the world's great teachers without leaving your home!

So I say, "Rah!" and "Fight on for Old SC!" to the USC Trojans. Its football players in their national championship halcyon days were known to have said, "Give us a faculty worthy of our team!"

"Yes, it's happening now!"

Chapter 7
THE TWILIGHT ZONE
1988-2004

NO WAY RETIRED!

In my 63rd year, the tempo of action at TRW started slowing down. Government money was tight and space vehicle procurement was slowed. Company belts were being tightened. My assignments became less interesting and exciting. I felt I was being pushed towards early retirement. Indeed, bonuses were being offered to those old-timers who opted for "early-out." Financially, retirement was entirely feasible. As a result of inheritances from my mother and my former nurse, as well as retirement packages from Aerojet, Iowa State, and now, TRW, our monthly income, supplemented by Social Security and IRA accounts, would remain about the same.

Given a helpful "push" by management, I left my beloved industry in July 1988. Getting into the semi-retirement mode was not the problem for me, as it was and still is for some of my buddies. Mostly, I missed the camaraderie of my friends at work, and the mental stimulation of either tackling a new problem or writing a new proposal. As mentioned earlier, I continued teaching at USC one night a week, for both semesters, until 1996, when I thought I had become technically obsolete. In early retirement, I occasionally helped my old TRW friend, Dr. Jim Wertz, start his new company, Microcosm, Inc., which is now prospering.

I began a daily regimen of writing in the morning and evening, taking a daily hot-tub/swim, followed by a walk, and sailing twice a week. My efforts to become a legitimate author have led me to write not only about my engineering and teaching careers, but also about my other love—music, along with some musings on current events. I continued, as I still do, to help out with my technical society's activities, and review the talks of speakers for their newsletter as well as giving lectures myself. I dare say, the hardest adjustment was my wife's—what with me being around the house every day! We also started traveling a lot. All this keeps me off the street and quite busy.

"Tilting at windmills" also continues, on both serious and nonsense levels. I am trying to get people interested in moving us into the "Hydrogen Economy," and trying to get NASA to develop a "lifeboat" for the International Space Station. Nobody listens!

N.I.H. ("NOT INVENTED HERE")—THE CREW RETURN VEHICLE METAMORPHOSES

An earlier story in this book, *"FIRST" In Space*, described a research and development program that took place in the early '60s. Its goal was to provide a small, compact, light, relatively inexpensive glider-like vehicle, which would act as a "lifeboat" in rescuing astronauts from a then-postulated orbiting space station. Forty years later, the ISS (International Space Station) was finally becoming a reality.

"Not Invented Here" is an insidious disease that infects large organizations. Nowhere in the modern world is this syndrome more apparent than in NASA, the main financeers of the ISS. This proprietary attitude all but precludes the adoption, or even consideration, of ideas that are not 100% home grown. I believe that the concept I conceived in the '60s is a victim of "NIH"—to the aerospace world's detriment. I will now tell you why.

In the early '90s, with the advent of ISS construction, NASA concluded that an emergency rescue vehicle was, indeed, necessary, and allotted funds to develop what they called the CRV (Crew Rescue Vehicle). NASA selected an old concept it had earlier developed, consisting of a stub-winged vehicle that would get them through the rigors of reentry into the atmosphere, followed by a parachute recovery. As the development proceeded, the system added more and more bells and whistles (I snidely suggested that it included hot and cold running flight attendants and a sauna). The cost overrun was so severe that NASA changed both its name and primary mission to the "Crew Return Vehicle," to replace the operational 3-person Russian Soyuz crew return vehicle. I objected to the new highly expensive approach by a flurry of "letters to the editor," and lobbied with NASA friends and foes. All to no avail. But in September 2001 an apparent program crossroad was reached. Again, I got into the act.

The main editorial in the September issue of *Aerospace America*, the monthly magazine of my technical society, the American Institute of Aeronautics and Astronautics, concerned the CRV program, and was entitled "Saving the CRV." It said, in part:

> *The station [i.e., the ISS] was designed to ultimately house a crew of seven. This would ensure sufficient hands to enable astronauts and cosmonauts to carry out scientific experiments on a regular basis, while at the same time seeing to the everyday station keeping.*
>
> *But it has always been understood that living on the space station was not a run-of-the-mill activity—one of the primary lessons learned from the Russian station Mir was that accidents can and will happen. If the seven crew members are going to be asked to take up occupancy, it is incumbent upon us to make sure they can get home quickly.*
>
> *NASA's Crew Return Vehicle is designed to serve this role—to get the crew home. Rather than docking a Russian Soyuz spacecraft, which can carry only 3 passengers, at the station, the CRV would be designed to carry six or seven. However, in this latest round of congressional budget talks, the space agency has recommended canceling the CRV program in an effort to reduce cost overruns.*

On page 10 of the same issue, a House of Representatives debate on whether to add another $275 million to the CRV program was described, with the outcome, apparently, leaving NASA to decide about which programs to cut or stretch out. These two articles prompted me to renew my battle, which I had earlier joined in an October 1997 "correspondence" with the editor, with copies sent to various NASA and key Congress dignitaries. My plaint then, in part, read: "Two articles in the June *Aerospace America* issue ('CRV investment offers safe return' and 'Mir: A

Health Report') followed by the real-time near-tragic accident on the Mir station, again lead to a conclusion that NASA considered and discarded about 10 years ago. NASA should now formulate a requirement for a true 'space lifeboat'—a minimal, very lightweight, throwaway rescue vehicle, several of which could be deployed at different station locations."

On October 8, 2001, seeing another opportunity to change history, I wrote to the editor, with appropriate copies sent to non-NASA bigwigs and the aerospace press:

"Both your editorial and a page 10 excerpt from the September, 2001 issue of *Aerospace America* noted, with apparent anguish, the possible demise of the CRV (Crew Return Vehicle, nee Crew RESCUE Vehicle). These articles imply that if the CRV is not developed, space station crew numbers will be forever limited to 3 persons, which, if true, is clearly not desirable.

"I, for the nth time, do not grieve the present CRV's probable demise. I gave my reasons for this stand previously in a "Correspondence" letter which appeared on p. 42 of the October, 1997 issue of AA. I still believe it is an ill-conceived, unnecessary, and Space Station utility-limiting concept which should be replaced by as many 1, 2, or 3-place inexpensive, light weight one-time-use-only minimally-equipped 'space life boats' as necessary to 'abandon ship' for all passengers aboard. Such true rescue vehicles, easily retrofit-attached, perhaps one-each, to all crew carrying station segments, allow the flexibility of returning one or two crew members at any time in case of, as an example, medical or family emergencies. Routine crew transfer would continue to be accomplished by the tried and true Soyuz capsule.

"Is a small, inexpensive, light weight, tightly packaged space life boat a real possibility? The answer was provided almost 40 years ago, and even then, the answer was "YES!", as I will now document. Attached to this letter, I have included copies of the salient documents which give credence to the technical aspects of such an undertaking.

"In 1962, inspired by the successful completion of an inflatable re-entry paraglider program we (Space-General Corp., a subsidiary of Aerojet-General) performed for NASA, I won a proposal from the USAF to design, build and test structural elements of an inflatable space station rescue vehicle. The one-crew version proposed weighed 1000 pounds and was stowed in a capsule that was about ten feet long and 5 feet in diameter. The containing capsule itself was an integral part of the heat protection system of the Nitrogen gas-inflated re-entry vehicle.

The final contract was for 1.3 million, and resulted in proving the feasibility of such a true space lifeboat. The idea was first presented to the aerospace world at the annual meeting of the AIAA predecessor organization, the IAS, in New York City in January, 1963. The paper, IAS # 63-35, was entitled 'Space Station Escape Vehicle.' During the time we were working on the Air Force contract, I proposed similar versions of the vehicle which carried additional passengers. But, by the time the USAF program was completed, it was apparent that there would be no space station in orbit by 1970 as had been predicted by Von Braun, and the concept was not continued. It was 30 years ahead of its time!

"Nevertheless, I believe such a concept should be reviewed in the light of today's more advanced technology and in view of the apparent sky-rocketing expenses being incurred by the CRV development. Such a minimal vehicle would, at relatively low cost, allow continued growth of the ISS, permitting additional crew members, and retaining the use of the Soyuz for crew changes.

"My hope is that the publication of this letter in *Aerospace America* will bring this alternative approach to the attention of those responsible. Note that I have previously made such a suggestion directly to NASA, but fear it fell in the NIH (Not Invented Here) wastebasket category."

Then came the next NASA vision, and a new road to salvation. In October 2002, in an article entitled "NASA Rethinks Plan for Space Station Lifeboat," *Space News* reported that delays in ISS completion allow time to re-evaluate its plan for acquiring 4-7 person CRVs. Some think that for a proverbial "few dollars more" (oh, sure!) they can develop a CTV (Crew Transfer Vehicle), which will not only bring crew back to Earth but can also transport them up to the ISS. This sounds suspiciously like a replacement for the Space Shuttle. Shortly thereafter, it too was abandoned in favor of a vehicle that could support lunar-return and manned Mars missions. Leave it up to good old NASA to find a more expensive and time consuming path to avoid what it must eventually face: That a true space lifeboat is what's really needed—in addition to a new space shuttle or exploration vehicle.

As I write this, in mid-2005, NASA is proceeding with a Crew Exploration Vehicle competition, which ten years from now would act as a replacement for the Space Shuttle—if such were still flying. The ISS is stuck with its maximum crew of 3 and a mission whose purpose was never very clear.

THE SWEET SURVEILLANT SCIENCE

In the beginning, I didn't know beans about the two surveillance-type instruments that Aerojet was expert in: microwave radiometers and infra-red (IR) systems. Both of these, along with radar and visual systems, had the capability of taking "pictures" of scenes on earth from space, and thus were very powerful from both a military and scientific standpoint. As a result of on-the-job training, I learned about the Aerojet instruments quickly, and was soon able to write proposals that kept these business areas growing. This experience, coupled with my earlier radar experience in the Navy during WW2, and later augmented by microwave radiometry and visual/IR experience at Hughes and TRW, allowed me to initiate, at Iowa State University, in 1973, the first known course in Remote Sensing Systems. I continued developing this course at USC from 1982 until my retirement from university teaching in 1996. Then, for a few more years, I presented 3-day remote sensing seminars all over the world.

By the sheer luck that has seemed to guide my wonderful 46-year jaunt through the aerospace firmament, I was again at the right place and time for getting into a brand new field—Remote Sensing. It was the early 1960s, and I was a systems engineer in the Space Systems Division of Aerojet/Azusa. They were doing pioneering work in both infrared (IR) and microwave radiometry sensors. The infrared work was being done under Project MIDAS, directed by John Jamison, and would lead to the development of the business end of the Defense Support Program Satellite. For many years, this satellite has been our first line of defense against missile attack. (It is also, incidentally, the world's best spotter of forest fires.) The microwave radiometry work, under the direction of Phil Caruso and his

RESOLUTION – A PRIMER
What can you see from space?

What Remote Sensing comes down to is a term called "resolution." This can be roughly defined as the distance between two identical objects when it first becomes apparent that there are, indeed, two distinct objects being viewed rather than one "blob." To humanize this, let's look down from space onto a large otherwise empty parking lot that has two identical sedans parked next to one another, two to 3 feet apart. The first problem is to detect that there is anything in the parking lot. This is called "detectability." If the resolution of the collecting system is better than, say, 20 feet, which is the approximate length of a side of the square area that the two cars make up, then the observation system has "detected" that there is "something" in the parking lot. Now, if the resolution gets to be less than two feet, the fact that there are two cars is "resolved." As the resolution further improves, you might be able to detect that the cars are identical and, at best, what make(s) they are. The latter would require additional knowledge that is called "ground truth" data, for example, the image of the identical car previously taken by the same recording system at approximately the same viewing angle and time of day. A comparison would then be made, just like fingerprint processing for matches.

AN INSTRUMENT PRIMER

A visible sensor—think of a camera—can only be effective in daylight and cannot see through clouds. An IR sensor can "take" pictures day and night and through normal cloud structure, but not with as good resolution as visible. A synthetic aperture radar can take pictures under all conditions, with resolution comparable to that of IR. Microwave Radiometers have poor resolution compared to the others, so they are used when viewing large, similar scenes, such as oceans or prairies. The aperture, or signal collecting device for these systems is either a mirror (visual and IR) or an antenna for radar and radiometer systems.

successor, Tal Falco, aimed to provide the first "modern" space-borne, electronically scanning imaging radiometer for the NASA/Goddard NIMBUS satellite program.

The modern era of sensing from space probably started in 1946, when photographs were taken from V-2 sounding rockets. The first real aperture (antenna size characterization) air-borne radar images were taken in 1950, and synthetic aperture images followed almost ten years later. Radar scatterometry (which conveyed return reflections only, without a picture) was done from the Skylab in 1973; two years later, NASA/Aerojet flew the Nimbus 5 electronically scanning microwave radiometer, which I had worked on in the 1960s. It was not until 1982 that the first truly modern imaging IR system was flown, in an early Landsat vehicle.

I have always been impressed by the ability of space-borne sensors to deliver timely information with great resolution and high sensitivity. These measurements have significantly enhanced our knowledge in the areas of agriculture and forestry and mineralogy and resource exploration, as well as Earth and planetary environments. We also know more about meteorology and weather, mapping and land use planning, safety systems, oceanography, and hydrology. In addition, these sensors will allow us to understand airway and traffic control. In the military world, remote sensing also has become the backbone of our defensive systems, including reconnaissance, missile detection and warning, satellite detection and tracking, and command and control systems.

We are now entering a decade in which the world has decided that surveillance with 1-meter (about 3 feet) resolution is acceptable. But such resolution is not even pushing the true capabilities of today's systems. For example, a 6-inch resolution in the visible range could be obtained from a 900-kilometer (about 540 miles) polar orbiting satellite, whose optical telescope has a 3-meter aperture (i.e., signal collecting mirror or antenna dish). A SR-71 Blackbird spy plane can resolve 1.3 inches with a 1-foot aperture at 70,000 feet. This means face recognition is possible! The progress in synthetic (i.e., electronically making the antenna much larger than its actual physical size) aperture radar systems has been equally impressive. With new technology allowing the detection of changes of a few hertz (i.e., cycles) in gigahertz (i.e., billions of cycles) carriers, the imaging capability of radar—during day and night and in good and bad weather—rivals that of cloudless-day visual images (pictures).

The beauty of all this is that data compression methods and both gigabit (i.e., billions of pieces of information) per second data processing and communication systems permit the collection and down linking of vast amounts of information, which cover large ground swaths in a relatively short time. Moreover, ground facilities can now convert and analyze this enormous data stream in almost real time. On the Earth and in the solar system, there is no longer anyplace to hide. "Big Brother" is watching YOU!

With remote sensing, we have conquered time and tide, as well as wind, water, and weather. We have found all the ships at sea, and have located ancient cities, scuds, under-ground waterways, holes in the ozone layer, and much more scientific information. These are truly impressive accomplishments, considering the relatively short time that remote sensing from space has been possible. And the best is yet to come!

THE HYDROGEN ECONOMY

As I write this, in March 2005, the price of a gallon of gasoline in California is at or is approaching an all-time high, and is increasing daily. Strangely, the citizens are not up in arms. A pervading attitude is that it's going to get worse, but what can we do?

The thing that really gets me is that, by taking one simple but brave step, our government could immediately cause the price of crude to drop to "give-away" figures, while ensuring a future of limitless energy supply, and a much cleaner environment. The step it must take is to announce to the world that the United States intends to be on the Hydrogen Economy within so many years (20, 30?), And to start moving in that direction.

I state this because I think that moving towards a world-wide hydrogen economy, made possible by using the clean, but not-yet developed, nuclear fusion process to produce the vast volumes of hydrogen needed to supplant oil, is the only and inevitable way to go for the future.

The Hydrogen Economy is a new way of life made possible when heat or energy is produced by combining hydrogen with oxygen. Both of these elements are available from the oceans of the world, and when they are combined, the process produces water—a closed cycle! We have found ways to produce electrical energy directly from this combination via a device called a "fuel cell." We also could use the combination in a combustion process that produces heat and/or thrust—as in a rocket engine. We could do this tomorrow if we had an inexpensive, environmentally secure way of producing vast quantities of hydrogen and oxygen. We already have an inexpensive way—nuclear fission—but the world, or at least the environmentalists in the USA, will not stand still for this "dirty" process. So, unless public opinion changes, we must rule it out.

What are the technical* problems in arriving at the seemingly ideal Hydrogen Economy? There are two big roadblocks: fusion energy must be developed—possibly by an effort the equal of the Manhattan Project—that produced nuclear fission, and a superior means must be found for storing, transporting, and utilizing gaseous hydrogen. Both of these technical problems are amenable to solution, and they only need significant money thrown at them. Fusion energy has been put on "hold"—barely kept alive at a few universities and national laboratories—for over 20 years. The problem here is more political than technical. People in high places do not want the world to be weaned from crude, nor do they want to be responsible for the havoc that will occur in countries whose very existence depends on oil. The hydrogen storage and transport problem probably already has a good solution—the storage of the gas in light metal as a hydride—releasable in gaseous form by heat and/or pressure. So, I think I've got a handle on the world of the future.*

But what I really want is to find out if I'm right, and if this is, indeed, the path to the future. I know I'm right about going on the Hydrogen Economy, that the oceans can provide all the hydrogen we will ever need, and that this is a replenishable process. What I'm not sure about is whether fusion is the only answer for its volume production. People have been working on other, less problematical, ways to produce hydrogen in vast quantities. Maybe there is another way.

There is a way to find out where we stand: We need to hold a world conference, wherein experts from all impacted areas gather to give papers on where they think we stand, and what we should do. I mean to include all factors—engineering, science, social, environmental, political, and economic—so that a picture of the future can evolve for evaluation. To this end, I have contacted universities with whom I have been associated either as a student or a faculty member, with the hope that I could interest them in sponsoring such a conference. I have sent a version of the below-copied letter successively to The University of Southern California, Iowa State University, the University of New Mexico, Cornell University, and New

---

* The anticipated social, political, economic, military, diplomatic problems are even more difficult to overcome than the technical!

York University. All have expressed interest, while sending me their "Dear Johns." It is true that I probably queried too low in the academic chain at USC and ISU—to the essentially money-starved Deans of Engineering. After learning that lesson, I went directly to school presidents, and wrote the following:

Feb.7, 2004

Dr. Jeffrey S. Lehman, President
Cornell University
Office of the President
300 Day Hall
Ithaca, NY, 14853

Dear Pres. Lehman:

I suggest that Cornell University undertake the establishment of an international conference on the very timely and critically important subject of "Fusion and the Hydrogen Economy." Because of its continuing interest in the fusion process and its interest in energy and socio-politico-economics, Cornell is well equipped to sponsor such a potentially earth-shaking investigation into the true state of the art and wither may go our nation's (and the world's) energy policy.

I am an Alumnus (BME, Sibley School, 1946) now retired from a professorship at USC and from industry (TRW). I am making this suggestion with the thought that such an undertaking would achieve much additional international prestige for Cornell, in addition to establishing the true feasibility of our independence from reliance on present non-renewable energy sources.

If, after reading more, you wish to probe further, I suggest you bring in an appropriate Vice President(s) for evaluation and implementation. I don't believe my proposal will get anywhere by forwarding it to the Dean of Engineering or to a Science Department unless it has your solid "go-ahead" endorsement plus promise of financial backing. In short, I'm asking for YOUR opinion, not the opinion of "experts" who may not have the overall vision or the horsepower to bring it off. Moreover, the scope of the proposed conference also requires participation by the economic, political, social science and other diverse Departments.

I am sharing the idea, to be further developed below, with you because I was a key participant in a similar important explorative conference in the '70s. It has occurred to me that the subject I now propose is equally or clearly more important to the world's future. Moreover (see back ground material, included), I think the time has come for such a similar ground-breaking event.

Before going into more detail on my proposal, I will first tell you more about the Conference alluded to above. It was the very impressive (and incongruous, for its venue) "Iceberg Utilization" Conference. Its purpose was to look at the feasibility of bringing fresh water to arid countries adjacent to open bodies of water. In the mid-70's, I—as Department Head of Aerospace Engineering at Iowa State—had a major role in this international conference, sponsored mainly by a US educated Saudi prince in charge of his country's water supply. The output of the conference, which attracted experts from all over the world, achieved worldwide attention and spawned several new enterprises. The "Iceberg Utilization" Conference covered the detection, selection, "lassoing," ocean towing (to arid countries), and "drinking from for a year" of Rhode Island-sized icebergs which regularly calve from the Ross Sea. It attracted about 250 aficionados from all over the "cold and arid" world and was organized under the aegis of the ISU VP/Research and conducted chiefly by the College of Engineering. It was an artistic, high impact success!

In my opinion, the time has come for a (similar) International Conference on the "Development of Fusion Energy and the Hydrogen Economy." The gist of the meeting would be to determine:
1) What's going on now in the world of fusion?

2) What's going on in the engineering aspects of reaction containment and the production of energy and the storage of hydrogen?

3) Research into other ways to produce large quantities of hydrogen?

4) What's the tie-in with the Hydrogen Economy? The environmental impact?

5) What infrastructures and funding are foreseen necessary?

6) Evaluation of the political/economic impact world-wide.

That's about it from me. I would hope that you might find a way, encouraged and funded by your Administration, to make it happen. If you are intrigued, I can provide you with the Proceedings (Pergamon Press) of the iceberg show and tell you how it played out on the ISU Campus.

Please let me know ASAP if you have personally read this and if you have an interest in pursuing it further. I feel very strongly about the need for such a conference and would push it elsewhere if you are not convinced.

Best regards, and much good fortune in your new job!

After a while, I received yet another "Dear John." Still undaunted, but no longer optimistic, I continue the search for a sponsor for this worthy concept. I have contacted my technical society, and they did look into it, but decided it was not their cup of tea. Next, I contacted DARPA, a government advanced research agency, and they too said it was beyond their purview. I still think it would be most impressively done under the aegis of a university. To this end, I asked a friend who lives in Princeton to try it out at Old Nassau, since Princeton is one place where some fusion research continues. No dice!

I remain convinced it is a conference whose time has come. Any takers?

A MAN WITHOUT A COMPANY

One by one, the Philistines have "made disappeared" all the organizations of which I am an alumnus. The final blows came quickly and mercilessly in June 2002, when the fabled Thompson-Ramos, TRW, came to its end-of-run. It was swallowed up by ever-expanding Northrop-Grumman, and became "N-G Space and Technology." TRW was my last place of regular employment. I retired from it in 1988. Earlier that year, the same voracious monster, Northrop-G., swallowed up the Azusa division of the Aerojet-General Corporation, for whom I had toiled, under one company name or another, for over 13 years, ending up as its Manager of European Operations in Paris. I am now an alumnus of—nothing! And this after a daily lifetime routine of 8-to-5 toiling at hard labor!

This excruciatingly sad story begins much earlier in time and my career. I don't count either my service in the Navy or work as an instructor at NYU while going through grad school as constituting career jobs. As recounted earlier, my first "real" job was at Sandia Corporation in Albuquerque. It was a secret nuclear weapons lab operated by the Western Electric Company, the same wonderful folks who used to be your sole supplier of telephones. I left there, the corporation still intact, in 1956. Somewhere in the '60s or '70s, it was absorbed into the government, and became what it remains today: the Sandia National Laboratory. Down one major place of employment.

I then migrated to what was then known as Convair/Pomona, a guided missiles merry maker, that soon became General Dynamics/Pomona, when Consolidated-Vultee was absorbed into G. D. Sometime in the late '80s or early '90s, the aerospace part of the General Dynamics empire disintegrated into a million pieces —the aeronautics part being bought by Lockheed-Martin, and my part, the missile end of the business, by Raytheon. Down two!

Sometime during this period, via an alumni magazine, I got the shocking and devastating news that NYU, from which I received master's and doctoral degrees, was eliminating both the whole damn College of Engineering and the entire Bronx campus. Can you imagine such a thing? The administration said it didn't want to dirty its hands any more by supporting a "trade school." It offered that Brooklyn Poly would nurture the campus-less NYU grads, if they required such nurture. I opted out. Down three!

In retrospect, I did my best work in the '50s at Sandia and in the '60s at Aerojet-General Corporation. At the latter, I first worked for the old-line company, but soon joined a wholly owned subsidiary, Space-General, and eventually became its Chief Engineer. But shortly after the corporate office sent me to Paris, it absorbed Space-General back into the mother company. Another organization—down four —shot out from under me!

The cycle was completed, as previously noted, when the Southern California branch of Aerojet became part of Northrop and, shortly afterward, the very same outfit bought up TRW. Down 5 and 6! I didn't even mention a brief, equally disastrous episode: In 1976-77, I spent a sabbatical year away from Iowa State at the then Hughes Space and Communications Group. It, too—down 7—recently was lately folded into the Boeing dynasty.

Except for Cornell University, which no doubt is open for gobbling up by the ever-expanding NYU; the University of New Mexico, which is doomed to being reduced to ashes in a now-being-plotted Iraqi-inspired "retributional" (for its support of atomic weapon development) nuclear attack; Iowa State University, which is always under threat of being absorbed into the University of Iowa; and my last part-time workplace, the University of Southern California—which inevitably will be absorbed by its arch rival, UCLA—they have no more homes to wrest away from me.

I am left wondering: Shall I go on attending the every-even-month gatherings of the Pasadena-based "Tejorea" Society (spell it backwards, you clever dogs!), watching the few remaining of my former colleagues growing older and more feeble at every meeting? Shall I continue convening with the TRA—the TRW

Retirees Association—whose yearly dues I have already coughed up. Or will we, like Moses' minions, be doomed to wander the streets of Redondo Beach for the next 40 years, muttering the following as if it were a mantra: Air Force begat Ramo-Wooldridge; R-W begat the Space Technology Laboratories; STL begat Thompson-Ramo-Wooldridge; TRW became Northrop-Grumman?

Who can I turn to? What shall I do? I AM A MAN WITHOUT A COMPANY!

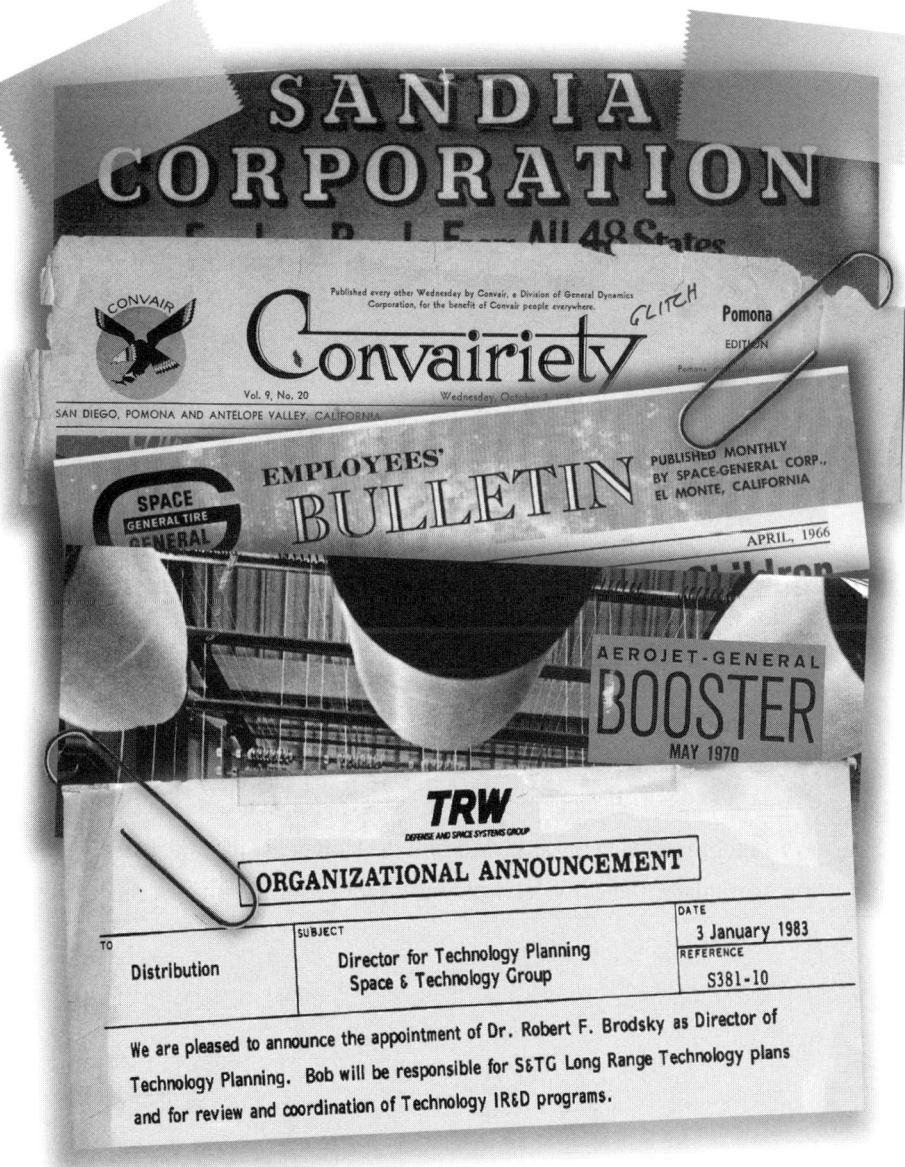

Chapter 8
SUMMING UP
2004-2005

LETTING GO—GRADUALLY

I knew that I wanted to be an engineer at age 10. It wasn't until I was 40, and hopelessly trapped in the profession, that I first wondered if I had made a good original decision. Maybe my mother was right? She always wanted me to be a "real" doctor. But, as a result of writing books and stories that cover my career in engineering and teaching, and recalling the many things that are unreported therein, as well as those that are, I am satisfied that, for me, the early choice was the right one.

As I proofread this book, I realize that I still have a myriad of stories—both technical and otherwise—to tell, and that at age 81, I'd better start writing faster and funnier. There are still many stories that I culled from an earlier version, which indeed, was part of a planned trilogy. There are additional stories that tell of the grandiose programs that ended up in failure or obscurity, such as the following: the NERVA nuclear rocket program, which was conducted jointly by Aerojet and the Los Alamos Scientific Laboratory and the adjunct RIFT (Reactor-in-Flight-Test) competition, which was cancelled; the successful, but dead end, SNAP 8 spaceborne 35 kilowatt nuclear power supply system, which came 30 years too soon; and the Aerojet AirTurboRocket, an air breathing engine that turned into a rocket when the altitude got too high for the atmosphere to sustain combustion—an engine that is sorely needed now.

I will also tell of some of the remarkable triumphs, like the saving of the TRW TDRS (Tracking and Data Relay Satellite) and the Hughes AsiaSat 3, both of which were delivered into doomed orbits by faulty launch vehicles and saved by sheer human ingenuity. And there are stories of latter day "expert witnessing" in patent infringement cases that are also intriguing.

Stand by for these and others. I ain't finished yet!

THE GLORY DAYS OF ENGINEERING

Engineering is an interesting and fun career. It was never better than in the halycon days of the '60s. In the early days of the space age, NASA, the Air Force, and the various big money "spook" agencies all desired to exploit the new frontier as quickly as humanly possible. They offered many "requests for proposals" to industry, which asked for outcomes that stretched the state of the art to extreme degrees. They bet on the "come," often asking for results that would be difficult to achieve, and possibly were known to be unachievable, without luck or a needed new invention. Industry, in turn, was more than happy to provide bids for these challenging quantum leaps, because it wanted to be at the cutting edge. Moreover its financial risk was limited.

This was so because the contracts let on these "chancy" projects were of a CPFF (cost-plus-fixed-fee) nature. This meant that industry bid on a job, quoting a certain price, and asked for an additional fee that was a fixed percentage of the quoted price: usually 2-3% negotiable. Receipt of the fee was a certainty, even if the job bombed, and it represented a clear profit. If additional project funds were later requested to continue, add on to, or strike out in a new direction, and if the customer agreed that such changes were justified, an additional negotiated fee—usually for a smaller percentage—came along with the "overrun" money. So, if you were clever enough to win a contract under these conditions, it was generally a win-win proposition for both you and the customer. You took the chance that you wouldn't foul up so badly that the customer would not do future business with you, and your reputation as an advanced research and development outfit went down the drain.

The atmosphere created by the CPFF way of doing business was exactly right for the times. It did permit the taking on of programs that people had heretofore only dreamed of. Even so, the success rate was remarkable, mostly because so many new ideas, analysis techniques, and simulation devices were suddenly available. Great strides were taken, even if many problems could be solved only by throwing great gobs of money at them.

So the competitions for new work were cutthroat, calling for vivid prose, big imaginations, and a few Hail Mary's thrown in for luck. But it was fun! It was an atmosphere I thrived in. In leading proposals—and I was very successful—I lived by the sacred words of my late Aerojet/Space-General boss, Charlie Roth: "Never say a job is difficult, never say we can possibly or probably do it. Just say we'll have it done next Tuesday, and be ready to punt!"

Recalling fondly those early days, and the rest, has led me to muse about engineering as an avocation that one could devote the best part of his or her life to. Should you, as a parent of a teenager, despite his or her gender, race, or denomination, recommend engineering and applied science as a career? You bet your sweet bippie you should!—IF your children meet two major criteria: They are good at and truly enjoy mathematics—attacking a word problem with the same zest as solving a crossword puzzle; and they are naturally curious—always wanting to know how things work. If they don't have both, or either, of these traits, they could still be engineers, and make a good living at it, but they'll never be GREAT engineers, and probably won't wake up every day anxious to go to work.

Engineering is good for the soul. It expands horizons; makes one want to see into the future; and stimulates one to be a part of the future one is personally carving out. Engineering hones your "people skills," just as much as your technical ones. Interaction and teamwork are necessities for success and are learned quickly. Knowing what your customer wants, and how much he will pay for it, are basic parts of the job. The old saw, "The customer is always right," is clearly the norm; yet you must know when even this axiom must be altered—by coaxing and cajolery. And you must write clearly and decisively, and be able to speak before large and small audiences with authority and clarity without, so to speak, "picking your nose." You must learn to respect other people's opinions, even when you know they are full of beans! In short, an engineer must be a well-rounded person, with the confidence to back decisive actions.

And what does an engineer with these attributes get? Well, for starters, lifetime employment at a pretty good, though clearly not sensational, rate of pay, and a lot of job satisfaction. And many perks come with the work, especially excellent medical and dental coverage plans for you and your family; a retirement pension that allows you to spend your latter years in relative comfort; paid travel to many far away and exotic places; and bargains and cut-rate tickets at the company store. And if you are very good and write papers, you might acquire a worldwide reputation, and a couple or 3 chances at the allowed "15 minutes of fame." But doggone it— it's just not as much fun as it used to be!

The air of engineering euphoria that I described earlier, brought on by the CPFF environment, went on for the better part of 10 years during the '60s. Buoyed up by the ongoing Apollo Moon program, it was a happy time in both the airplane and space sectors. But every now and then the public got a glimpse of what was going on in this happy military-industrial complex that "Ike" (President Eisenhower) first denoted. Scandals like $100 screwdrivers and $500 toilet seats leaked out, until Congress started paying attention. Why not "fixed price" (FP) it said? And soon the pendulum started to swing. But, of course, it soon swung too far.

As a result, the day of widespread CPFF R&D contacts waned greatly, to the minority position it holds today. Instead, contractors are forced to bid a fixed price that they agree they will not exceed, even if they have to finish the job using their own money. In return, they are allowed a higher profit margin to help cover their much greater risk, especially in undertakings that normally would be naturals for CPFF agreements. The mind-set in bidding FP is much different from bidding CPFF. You can no longer depend on the customer to bail you out if you get in trouble. It is a different ball game, and one that I think is, in many cases, to the detriment of the country. The government, in some key developments, now asks for FP bids on programs that are clearly of a speculative nature. The customer knows it, and the contractors know it and complain, but both are cowed by a recalcitrant Congress.

When this new regimen started to set in, the results for both old-timer, advanced engineers and engineering were just short of being disastrous. The wind was taken out of their sails. On the face of it, this very logical edict was saying, "Let's only buy what we know we can achieve. Let's no longer bet on 'the come'." They were overly cautious. The newer generation of engineers, who were not active in the CPFF days, didn't know they were missing mind-expanding opportunities.

For engineers under this new regime, the challenge is to find an approach, based on proven technology, that no one could doubt would work. The soaring flights of innovative minds were suddenly grounded. In place are good, but uninspired and expensive, solutions to problems that produce an inch, not a yard, of progress. New organizations called "Risk Management" have sprung up, and are manned by engineers who, heretofore, were doing advanced design work. And engineering is no longer the fun it used to be.

Yes, the swashbuckling days are over, except for a few crucial CPFF jobs that slip under the door—usually ones so cloaked in secrecy that you never hear about them. The change took the excitement of going to work every day away from many of our best innovators. It certainly made it easier for me to retire when my time came. "Fixed price" is clearly a mixed blessing. It signifies that we have reached a certain stage of engineering maturity, but methinks we are paying a high price in progress for it. Still and all, compared to most professions that I'm aware of, I'd continue to put engineering high on the list!

Interoffice Correspondence
TRW Space & Technology Group

Subject: An Offer You Can't Refuse
Date: February 21, 1984
From: Your CAPO

To: Group Staff AAP Committee

cc:

Location/Phone: R5/1031 61824

At great savings to the women and minorities of the world, we have chartered the fabled sloop, "Poulet de la Mer" and its equally noted British Captain, R. Foxroy Broadbeam, for a series of late Wednesday afternoon group staff AAP committee meetings. Alas, the sloop, a magnificent Catalina 27, only holds six committee persons comfortably, so a number of cruises will be necessary to get the work of the committee done. Reservations will be taken by contacting the Captain's executive officer, Cheryl, at x61824. The boat leaves from slip B-9 at the King Harbor Marina (see map) promptly at the arranged times. Tennis shoes are de rigueur. The Captain provides ice and a grim visage - you bring the committee business!

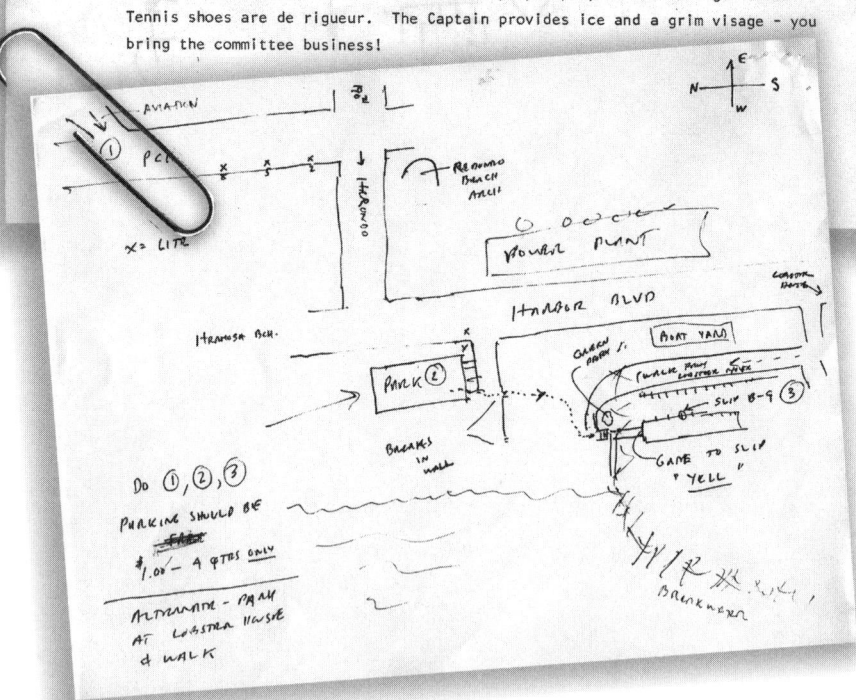

All work and no play makes Jack a dull boy!

THE LAST HURRAH
or . . .TRAVELING AIN'T FUN LIKE IT USED TO WAS

Professor Mike Gruntman, head of the Astronautics and Space Science Division at USC, suggested that I submit a paper to the annual Aerospace Sciences Conference of our technical society. He thought that an account of the pioneering work that I had done in the '60s, on emergency escape from a space station, would be of interest to aerospace historians.

I mulled over the idea, thought it might be good for my ego, and believed that, if it were accepted, I would probably be acclaimed—at pushing 80—as the oldest speaker to ever address this august body. I also remembered, from many previous treks to the conference when I was a young buck, the most excellent Basque restaurant in downtown Reno, which clearly warranted a revisit. So, I submitted a paper "abstract" to a historical session, and it was accepted. Thus was born, for me and my wife, who shared my fond memories of the Basque, a trip from hell!

The jaunt to Reno in January 2005 was strictly an ego trip—and an expensive one, at that. Earlier, since its venue was in nearby Long Beach, I had submitted a technical paper to the annual conference, "SPACE 2003." As discussed in chapters 4 and 7, it dealt with the design of a unique emergency rescue vehicle that I had "invented" in 1962, to bring back crewmembers from a failing space station. I had spent considerable effort, starting in the late '80s, trying to convince NASA that such a device was needed for the International Space Station (ISS)—to no avail. Surprisingly, the paper was not accepted. A year later, further discouraged by the new U.S. policy, which had all but abandoned future development of the ISS, I reluctantly decided to give up the fight to legitimize the space lifeboat. So this final paper was to be the last hurrah for the space lifeboat.

AIAA Dinner Meeting

"ESCAPE FROM THE SPACE STATION"
Speaker: Dr. Robert F. (Bob) Brodsky

In the early 60's, Wernher Von Braun confidently predicted that the United States would have a manned space station in place by 1970. This statement prompted this month's speaker, Bob Brodsky, to propose a "lifeboat" vehicle to rescue station crew in case of dire emergencies. So confident were people of Von Braun's prediction, that the USAF funded, in 1963, a partial space station escape vehicle development for an amount that eventually reached $1.3 million. A flurry of international publicity followed his January, 1963 reading of his initial paper to the annual meeting of the "Institute of the Aeronautical Sciences," one of the AIAA predecessor organizations.

Emphasizing that his design is more pertinent and equally viable today than it was 40 years ago, Brodsky has been arguing that the present space station, like an ocean liner, needs an emergency one-time-use-only rescue "lifeboat, in addition to the "Orbital Space Plane" crew delivery and return vehicle program that NASA has recently embarked on. Brodsky's talk will present this message and give details of the considerable body of work that was accomplished in the heady flush of the beginnings of the space age in the 60's. The design featured single crew lifeboats that weighed less than 1,000 lbs and were packed in a 3' x 10' cylinder. His historical work will be addressed at the dinner and will be featured in an article to appear in the October, 2003 issue of the Smithsonian "Air & Space" magazine.

DO NOT MISS THIS EVENT!
Seating will be limited so get your reservations in early. It is important that if you expect to have dinner, you must call in your reservation by the RSVP date.

$30.00* Regular $25.00* Young Professionals and Military $15.00* Students

$5.00 Attendee Only – No Meal

Show your AIAA card for $5.00 discount

(Corporate tables are available. Call or email Dr. Robert Conger,
310-726-4100, rconger@smad.com for more information)

You must RSVP by Friday, September 12 to ensure a meal reservation. Call the AIAA reservation line at (800) 683-AIAA (2422) or via e-mail to christinak@aiaa.org

CANCELLATION POLICY
If you cannot attend, please cancel your registration within 48 hours of the event by sending an email to christinak@aiaa.org and mentioning your cancellation. If you do not cancel your reservation within 48 hours of the event, we expect you to submit appropriate payment to :
AIAA, 2221 Rosecrans Avenue, Suite 227, El Segundo, CA 90245

Our flight to Reno was to leave LAX at 10 a.m. on Monday. Our Santa Fe daughter, her Santa Barbara son, and his lady friend had been visiting us, prior to winging it to Hawaii on an earlier 9 a.m. flight. On Sunday evening, I arranged with the usually reliable Yellow Cab Company for a 5-passenger van to pick us all up at 6:45 a.m. next morning. The dispatcher said the driver would call us as he approached our pickup point. Previous experience was that they usually arrived about 15 minutes early, so when no one appeared by 6:50, I called in, slightly panicky. It was a mind-blowing revelation to find that she had never heard of me, but that she bet she could "have a cab there in 10 to 15 minutes." Thinking quickly, I asked, "Do you have any cabs at the Crowne Plaza Hotel," which is directly across a small street that runs between our buildings. She allowed that there, indeed, were two vehicles now stationed in front of the hotel. "Fine," I said, "please call one and tell him to come by immediately!" "OK, they'll be there in 10 to 15 minutes." Being no longer able to deal with incompetence, I told her to "forget it." I plodded across the street in the rain, and commandeered a waiting cab at the hotel front entrance. The first crisis was over, but I was wet and out of breath—and short of temper!

At the airport, fortunately with plenty of preflight time, we ran into the vicissitudes of modern technology. You will see that this eventually led to an attack of apoplexy on my part, with an accompanying first notice divorce warning from my dear wife. Alaska Airlines, our carrier of choice, has instituted a mechanized way to get its customers started on the long process of checking in baggage and getting boarding passes. Innocently enough, it started with a supposedly user-friendly computer screen and keyboard. My wife carries the tickets, so she took on the machine. The first thing it asks, once you get it energized, is "Confirmation Number, please." Ay, there's the rub, since there wasn't a confirmation number on our electronically derived ticket. And try as we both might, by calling for "Help" on the screen and other futile stabs, we could not get past the first barrier. Thus stymied, we were not able to move on the open agent stations that were ready to immediately service those who were able to deal with the machine.

I then moved both our baggage and my heightened blood-pressured body into the long line of those who, also, had not been able to use the machine, and were waiting to be processed by a real person. In the meantime, my dear wife had found a minion lurking around the intransigent machine, who told her that if you didn't have a confirmation number, you merely had to type in your name. She beckoned me to abandon my position in line and join her at an open machine. Giving up our place in the queue, I joined her, while we tried the new formula. And, try as we might, we could not bring up an icon that asked for our name. Not being noted for patience, the "evil twin within me" began to mumble under his breath—speaking unkind words about our airline of choice.

But, before an explosion occurred, we spotted another floating minion, whose job appeared to be of an advisory nature, since he appeared to be "helping" another hapless soul at the adjacent machine. I signaled audibly for help, and he replied, "Wait a sec, it's easy." With that, I roared, "Easy!! Fer Chrissakes, I've been here 15 minutes, and even with my PhD in Engineering, I can't figure this dumb sumbitch out!" With that, he disdainfully declined to help me, noting that rudeness doesn't count, and that he "wasn't being paid to take this kind of treatment!" So I got back at the new end of the line, suggesting to my wife that she politely try him out. But he was too slick for her, and told her that he knew "she was with the madman." Eventually, we got to the front of the line and checked in, noting to the agent that we could be contacted at the Reno Hilton. Blood pressures back down to their normal highs, we arrived in Reno, determined to use Alaska again only if it were the last plane out at Armageddon.

The meeting itself went well: I schmoozed with some old industry friends and university colleagues; we enjoyed the Basque restaurant, despite the cold, slippery, slushy sidewalks in downtown Reno; and my talk went well, and was well attended. Two grad students, looking into lifeboats, collared me after my talk, to discuss their thesis work in space rescue. So my ego was sated, although, in retrospect, not $800 worth, and especially not after the final Alaska Airline affront.

We were to return to LA via a noon plane, and reached the airport at 10 to begin the check-in process on the evil machine. But this time we were armed with our confirmation number, which the agent at LAX had provided. We started the process and all went well, until it asked "Destination?" We typed in "LAX" and it asked "Destination?" We typed in "Los Angeles" and it, again, asked "Destination?" I felt the tightness gathering around my chest. Fortunately the nearby agent called out to find the problem. She told us, "Oh, didn't you know the flight had been cancelled?" No, we didn't, although we would have known had we been at home, since that's where they called to inform our answering machine of the change in schedule. They did put us on a 4 p.m. flight the same day, so we went back to the hotel, stewed, and spent more money. Altogether, an ugly experience.

To cap it off, about an hour before my talk, I had run into a very old (from the '60s) friend, and we caught up with each other's adventures in life. I casually remarked to him that, at almost 80, I would probably be the oldest living person to give a paper at this highly technical meeting. He replied, "I gave my talk yesterday, and I'm 83!" In true Linus fashion, "Rats," I said.

But, my friends, there's a happy ending to all this. I wrote an explicit letter to the president of Alaska Airlines expressing our displeasure with our journey, while making suggestions as to how it might improve things. I did this just to vent my spleen, never expecting an answer. About a month later, I received a very conciliatory letter from a customer-service lady, who said the president had read my letter with interest and dismay, and directed her to credit our account with the full roundtrip cost of both of our tickets!

Postscript: In January '06, we made a round trip from LAX to Puerto Vallarta on Alaska. All went swimmingly. No machines were involved.

THE GREAT EVENTS OF THE LAST CENTURY

The past 100 years were marked by giant strides in the arts and sciences. But no field, with the possible exception of medicine and nuclear energy, made as remarkable an impact on our everyday lives as did the progress in aerospace engineering and science. As a new century is getting under way, it is fitting to remember and record the most important of these giant strides, and try to give proper emphasis to the ones that represented giant break-throughs.

Such an undertaking is bound to be controversial, even if compiled by a committee of expert historians. Unanimous "greatness" ratings would be almost impossible to come by. I unsuccessfully tried to synthesize a similar article 25 years ago, with inputs from noted aerospace writers/historians, pertinent technical committees, aerospace educators, and magazine editors. At that time, aiming for a "top ten" list that covered the then past 50 years, we could not get anywhere near to unanimity and gave up. So, this time, I decided to go it alone, retaining the wisdom of the earlier survey, while seeking and incorporating guidance from knowledgeable e-mail and local friends. "And the devil take the hindmost!"

To start the list, it's hard to see how anyone would not agree that the Apollo Moon-landing program joins the Wright Brothers' feat as the two head-and-shoulders-above-the-rest aerospace headline events of the 20th Century. Next in line, but already entering the controversial, might be Lindberg's solo flight across the Atlantic, which led to regularly scheduled, transoceanic travel; Billy Mitchell's bombing effectiveness demonstration, which forever changed war fighting; the advent of the DC-3 "airliner," which initiated profitable commercial air travel; the development of ballistic missiles, which changed the balance of power and opened up space; the rise of communication satellites, which have become an important agent in unifying the world; and the Mars instrument package landing, which heralded the intimately detailed exploration of the solar system and the universe. But from there on in, it's anybody and everybody's opinion.

In the following listing of leading events, I have included all of the many milestone events that my collaborators and I have agreed upon. Further, I have emphasized those that I think are truly outstanding. I have not tried to list events in historical sequence, leaving such an exercise to true historians. The great electronic (e.g., solid state devices, Klystron tube/communications systems, radar, and remote sensing instrumentation) and atomic energy breakthroughs were not included since they were deemed *not* strictly aerospace-related.

Here, then, are the 15 major milestone categories, with the individual events deemed most important both italicized and bold-faced:

THE GREAT AEROSPACE EVENTS OF THE 20TH CENTURY

1. THE ACHIEVEMENT OF HEAVIER-THAN-AIR FLIGHT
The ***Wright Brothers*** develop airfoil and airscrew theory, lightweight engine technology, stability and control theory, wind tunnel and analysis techniques; they ***build and fly the first successful heavier-than-aircraft, the "Flyer,"*** in 1903.

2. THE EARLY DEVELOPMENT OF AIRCRAFT
Biplanes, seaplanes, amphibians, monoplanes; the work of ***pioneers Curtiss, Loening, Douglas, Martin & Northrop***; barnstorming, aerial warfare, mail delivery systems; tricycle landing gear; cantilever wings, air cooled engines; private aviation (Stinsons, Beeches, Cubs, Cessnas); the 'Smilin' Jack' comic strip captures kid's imaginations; ***Doolittle demonstrates "blind landing"***; the Supermarines; propeller driven fighter and light bomber aircraft; the GB Super Sportster.

3. GREAT EARLY EXPLORATORY FLIGHTS

Bleriot crosses the Channel; the Douglas World Cruisers global flights; *trans-Atlantic flights (e.g. NC-4, Lindberg); North and South Polar flights* (Adm. Byrd); London-Australia non-stop flight; famous transcontinental flights (Bergdoll, *Hughes*); the Wiley Post/Will Rogers jaunts; the Lindbergs' travels, as documented in *Listen the Wind*; Amelia Earhart's final flight; Picard's balloons; the development of parachutes for safety, bomb and airplane deceleration, and heavy cargo soft landing, stratospheric parachute jumps.

4. THE ARRIVAL OF LIGHTER-THAN-AIR CRAFT

The golden age of the rigid LTAs: The Graf Zeppelin and its successors, the Los Angeles, Shenandoah, Akron, Macon, and the *ill-fated Hindenburg*. Later utilization of non-rigid blimps and propane-fired recreational balloons; high altitude balloons.

5. RISE OF COMMERCIAL AVIATION & HIGH SPEED PROPELLER-DRIVEN AIRCRAFT

Metal and monocoque construction; the Ford Trimotor; *the DC-2/DC3's revolutionize transcontinental travel*; the ensuing DC-X series; Eddie Rickenbacker starts Eastern Airlines; *Gen. Billy Mitchell* proves *efficacy of strategic bombers; Doolittle's raid on Tokyo*; the XB-15, *B-17 Flying Fortress*, B-29, B-50, B-24, B-36; the Dornier DO-X; the Sikorsky and Martin flying boats *(China Clippers)*; the famous WW-II fighter planes; the flights of *Enola Gay* and TWA 800.

6. V.T.O.L. AIRCRAFT

De La Cierva/*Sikorsky development of the helicopter*; Kellett/Pitcairn development of the *autogyro* with *mail delivery in Philadelphia*; the Ryan/Convair tail sitters; XV-12 to Osprey; the Piasecki "Jeep."

A postcard commemorates the Dornier Do-X's first flight on July 13, 1929. The flying b[oat's] disappointing performance led to a major engine change.

Dornier DO-X, circa 1929. I watched it fly by the beach at Atlantic City in the early '30s. The 12 engines made a lot of noise. From Aviation History, March 2001, Primedia Inc.)

7. THE ARRIVAL OF JET PROPULSION

Inventors Von Ohain and Whittle build gas turbines; the first jet aircraft, the ME-262; the Ryan Fireball; subsonic fighters (culminating in the F-86), bombers (B-47, B-52) and commercial aircraft (e.g. the ***Boeing 707***); the Skunk works and the *U-2*.

8. THE AGE OF SUPERSONIC FLIGHT

The X-1 and Chuck Yeager; the X-series culminating in the ***X-15***; the YF-12 and the F-100 series of fighters/fighter bombers; the ***SR-71 Black Bird (Kelly Johnson)***; B-58 Hustler; the ***Concorde SST.***

9. THE AGE OF MISSILES/ROCKETS

Early ***liquid rocket development by Goddard***: The British and American Rocket Societies and the Buck Rogers comic strip pave the way to space; the V-2 and the Von Braun saga; ***development*** *of ballistic missile vehicles in the U.S. (Atlas, Titan, Delta) and USSR;* short range missiles; surface-to-air and air-to-air missiles (Nikes, Falcons, Terriers, Side-winders, etc.); Minuteman; *the sounding rockets (Aerobee).*

10. DISCOVERING NEAR SPACE

The Vostok/*Sputnik*, Explorer, and Vanguard initiate space utility; the development of launch vehicles via the *IRBMs* (Thor) and ICBMs; the IGY; restartable (in space) upper stages—the *AbleStar*; the early scientific satellites discover near earth phenomena, especially the *Van Allen belts* and the solar wind; weather, communication, nuclear explosion detection, earth resources, reconnaissance, intelligence, early warning, scientific, oceanographic, and navigation (culminating in the GPS) satellites are born; *Syncom achieves geosynchronous orbit (as predicated by Clarke)*; observatory satellites look outward (Hubble, Compton, Chandra, etc.).

11. MANNED SPACE FLIGHT

Laika's flight and the U.S. chimps; *Gagarin's first flight*; Shepard and Glenn flights; the Mercury and Gemini programs; first EVAs; the Ranger and Surveyor precursor programs; *the Apollo Program and Armstrong's Moon landing*; O'Neil's space colonies; the Dynasoar, the Skylab (repair in space); the *Space Shuttle*, the Apollo-Soyuz in-flight connection; Soyuz/Mir space stations; the ISS begins as the century ends.

12. DISCOVERING THE PLANETS

The first deep space visitors, the Pioneer and Explorer; *Viking lands on Mars*; Mariner—and Pioneer—Venus and Russian Venus lander; the continuing search for extra-terrestrial life; *Voyager & Pioneers leave the solar system*; Galileo; Asteroid exploration; the slingshot trajectory approach by JPL.

13. MANNED AND SOLAR-POWERED FLIGHT
The long-sought prize; *McCready and the Gossamer Condor*; the flight over the English Channel; the solar powered flight in Greece; the high altitude satellite-substitute robot is designed.

14. ROUND-THE-WORLD NON-STOP FLIGHTS
The B-52, with in-flight refueling, circles the globe; *the remarkable flight of the Voyager with Dick Rutan/Jeana Yeager*; the globe encircling balloon flight.

15. THE GREAT TECHNICAL/MANAGEMENT ADVANCES
The formation of NASA, ESRO/ESA; the development of launch sites provides access and operational procedures to conquer space; the management of the Apollo program: the rise of the "ilities" (reliability, value engineering, PERT, risk management); the invention of the fuel cell and solar cells to facilitate manned space flight; *the invention of the transonic wind tunnel by Stack et al*; solutions to the reentry heating problem, and the great enabling technical and analytical advances (*Tsiolkovsky, Homann formulate modern astronautics*); Singer proposes spin stabilization of satellites, the development of potential theory to solve low speed flow problems, *Jones' cross-flow theory*, leading to simplified swept-back wing flow analysis, *Whitcomb's "Area Rule"* facilitates the transition from subsonic to supersonic flight, "finite element" structural analysis theory, the rise of *computational fluid dynamics*, CAD/CAM, the conquering of boundary layer and turbulent flows, "smart structures", and *stealth*.

The above events have been limited to the aerospace field. The advent of the transistor, which enabled some of the listed innovations; the atomic bomb; the GPS, penicillin, and the defeat of polio, and organ transplants; and personal computers are, in my opinion, on a par with the development of heavier-than-air craft and the Moon landing as the outstanding all-category scientific/engineering movers and shakers of the past century. You will also see that the aerospace-related listings above are heavily slanted towards American achievements, and I am sure that Europeans and others might have a somewhat different viewpoint. When you think about it, you will recognize that the last 100 years gave birth to at least 98% of the total all-time aerospace advances. It is difficult to see how this new 21st century can achieve breakthroughs to equal the tremendous quantum-leap progress of the past century. We can foresee major happenings such as manned Mars missions, Single-Stage-To-Orbit vehicles, flying automobiles; hydrogen-fuelled aircraft; an L-5 space colony; a loving couple becoming the first members of the "250 mile high" club to go public; and, hopefully, the huge breakthrough in propulsion that might allow us to better explore our universe. But my guess is that, as in the last 20 years, progress in aerospace will be incremental with only a few headliners akin to the many high-impact milestones that marked the 20th century.

But don't despair. I have a notorious history of under-estimation; the above prognostication should probably be relegated to the "famous last words" trash bin. Finally, as we finished the old century, the initiation of the assembly of the International Space Station indicates what seminal event may lead off a similar listing 100 years from now. But clearly, the 20th century was a great one for aerospace endeavors. Long may it wave!

The old "Perfesser" with a dented horn.

INDEX

Abbott, Ira, 31
AEC (Atomic Energy Comm.), 1, 23, 28, 35, 37, 41, 60, 64
Aerobee, 5, 88, 105, 106
Aerojet-General, 1, 2, 74, 81, 82, 84, 85, 87, 106, 110-112, 114, 115, 118, 119, 121-123, 126, 129, 140, 143, 177, 180, 182, 184, 191, 192, 195, 197
Aerospace Sciences Conf., 201
Affirmative Action, 159, 169, 171
AF CRL (Cambridge Res. Lab.), 88, 106
Air Cushioned Vehicles, 114-117
AIAA (Amer. Inst. Aeron. & Astron.), 1, 146, 148, 155, 179, 181
Al Faisal, Mohammed, 130
Analog computers, 70-73
APL (Applied Physics Lab), 70, 74
ASAE (Amer. Soc. Aerospace Educ.), 144
ASEE (Amer. Soc. Engr. Educ.), 144
Barsh, Max, 126
Beech Aircraft, 141-143
Bell Labs, 28
Boylan, Dave, 153
BRL (Ballistics Research Lab.), 45
Bumblebee family, 70
Caruso, Phil, 182
Charters, Alex, 31, 46-48
Convair/Pomona, 1, 69, 70, 71, 74, 76, 81, 83, 122, 191
Cornell U., x, 4, 8, 11, 15, 32, 62-64, 66, 187, 188, 192
Cornell Aero. Lab., 62
CPFF, 196
Crew Rescue Vehicle (CRV), 98, 178-181
Cronvich, Les, 74
Crowder, Ken, 39

CWT (S.C. Co-op Wind Tunnel), 62
Darakji, Dave, 119
De Los Santos, Soc., 15
Douglas DC-3, 12
Draper, Eaton, 34
DSP (Defense Support Prog.), 122, 123, 127
Eimer, Fred, 86
ENIAC, 15, 18, 19
Erickson, Ken, 34, 62
Fat Man, 28
Fink, Dan, 62
Fixed price, 198
Fletcher, James, 82, 87
FLTSATCOM Satellite, 164
4925th, Strat. Bomb. Squad., 23, 26
Froehlich, Jack, 82, 87
Fuel cell, 186
Fusion, 186
Glennon, Nan, 169-170
Great Aerospace Events, 206
Grumman, Leroy, 11
Gruntman, Mike, 149, 150, 201
Gutierrez, Leo, 60
Hansche, George, 34
Harkin, Tom, 129
Heinemann, Ed, 39, 41, 45
Hill, Paul, 31, 45
Hughes Aircraft, 2, 157
Husseiny, A. A., 131
Iceberg Transport. Comp., 134
IMP (Inflatable Micromet. Paraglider), 94
Iowa State U. (ISU), ix, 2, 37, 129-131, 135, 136, 140, 145, 153, 154, 157, 182, 187-190, 192
Isaacs, John, 130

IR (Infra-red), 182
IRAD (Indep. Res. & Dev), 168
IUS (Inert. Upper Stage), 162
Jamison, John, 182
Keville, Bud, 94
Kimball, Dan, 118
Kinard, Bill, 13
King Hassan of Morocco, 118
Klemin, Aleander, 7, 10-13
Kraft, Al, 135
LDEF, 137
LEASAT, 160
Little Boy, 28
Loening, Grover, IX, 11
Los Alamos Sci. Lab. (LASL), 23, 28, 60
Laitone, E.V., 55
Lake Bemidji Test Base, 23
Levy, Larry, 60
Ludloff, J.F., 15
Mattson, Axel, 65
MARK (Mk) 3 A-Bomb, 29
Mk 4 A-Bomb, 30, 31
Mk 5 A-Bomb, 34-41
Mk 6 A-Bomb, 33
Mk 7 A-Bomb, 42, 43-49
Mk 8 A-Bomb, 28, 43
Mk 12 A-Bomb, 49
Mk 13 A-Bomb, 33
Mk (TX) 17 H-Bomb, 60-65
Martin, Glenn, 11
Merritt, Mel, 23, 24
Metal hydrides, 187
Michaelis, John, 50
Microcosm, Inc., 179
MIDAS Program, 182
Migdal, Nick, 108

Miller, Jack, 100, 103
Morrison, Al, 100,103
Morton, Jelly Roll, 144
Muscatine Air Line, 141-142
NTU (Nat'l Tech. Univ), 174
New York University (NYU), 4, 7, 10, 15, 129, 152, 191, 192
NOL Daingerfield wind tunnel, 73
Northrop, Jack, 31
Northrop-Grumman, 123, 191, 193
OFO (Orbiting Frog Otolith satellite), 90
OV-3 Satellite, 89-97
PARD (Pilotless Aircraft Research Div.), 45
Petersen, Robt. P (Pete)., 21
Poor, Charles. III, 45,111
Pope, Alan, 37-40
'Q' clearance (AEC), 24
Rancho Los Amigos, 103, 104
Reid, H.J.E., 64
Remote sensing, 122, 130, 131, 132, 139, 146, 154, 155, 159, 172, 182-185, 192, 207
Resolution, 183, 184
RIFT, 195
Risk Management, 199
Ritland, Osmund J., 34, 35, 39
Rogallo, Frank, 94
Rohr Aircraft, 120-121
Roth, Charley, 81, 84, 197
Rowe, Paul, 37, 62, 129
Salton Sea Test Base, 23, 35
Samuel, M. Abel, 114-116
Sandia Corp., 1, 21-23, 28, 57, 58, 64, 66, 69, 70, 81, 129, 191, 192
Saturn S-2 Stage, 83-87
Schairer, George, 31, 35

Shreve, Jim, 24
Shuttle Bus, 160
Sibila, Al, 45
Sommer, Howard, 122
SNAP-8, 195
Space-General, 1, 81-84, 88, 89, 100, 106, 110, 111, 118, 119, 180, 192, 197
Space Station Escape Vehicle, 92-99
SPAM (Hormel), 136
Spangler, Gene, 164-166
Stack, John, 65
Swanson, Ted, 150
Surveyor Moon Lander, 100
Syphilis test device, 111-113
Tartar missile, 69, 74-78
TDRS satellite, 195
Teller, Edmund, 60
Terrier missile, 69, 71-74
Thompson, Floyd, 64
Truax, Bob, 106
TRW, x, 2, 34, 122, 123, 154, 156, 157, 159, 160, 162, 164, 167-70, 172, 177, 182, 188, 191, 192
University of Southern California (USC), 2, 150, 154, 157, 159, 169, 172-74, 177, 182, 187, 188, 192, 201
USS Norton Sound, 106
Van Every, Kermit, 45
Vaughn, Harold, 44
Velton, Ed, 69
Vlasek, Frank, 61
Von Braun, Wernher, 92
Wallops Island Test Base, 106
Wang, C.T., 15
Wertz, Jim, 179
Western Electric Co., 28, 191
White Sands Proving Ground, 106
Wilson, 'T', 35
Wings Club, 11
Wright, Orville, 11
Young, Bob, 85, 148
Zonars, 'Zip', 32